Radioactivity: A Very Short Introduction

Claudio Tuniz

RADIOACTIVITY

A Very Short Introduction

Acknowledgements

I am greatly indebted to many colleagues for their critical review of the manuscript and suggestions for improvements. In particular, I would like to thank Alessandro de Angelis, Abdelkrim Aoudia, Marina Cobal, Paolo Creminelli, Fabio de Guarrini, Colin Groves, Cheryl Jones, Juergen Kupitz, Francesco Longo, Roberto Macchiarelli, Mariarosa Malisan, Francesca Matteucci, Colin Murray-Wallace, Michele Pipan, Nevio Pugliese, Holger Rogner, Barbara Stenni, Filippo Terrasi, Patrizia Tiberi Vipraio, Alessandro Tuniz, and Kevin Varvell.

Contents

Radioactivity

Introduction

You cannot hide from radioactivity.

Even the book you are holding is slightly radioactive. But relax and read on: radiation effects to your eyes and body are below the maximum recommended by the International Commission on Radiological Protection. You should be even less worried if you are reading an e-book.

There are more serious risks, such as those emerging from the radioactive gas radon (radon-222), a product of natural uranium that can invade your house from the soil beneath it. Building materials, including granite, concrete, and bricks, can emit significant volumes of radon. Radioactive gas does not have any odour, colour, or flavour, and can accumulate, undetected, to levels that can be dangerous to your health. In many countries, this is the second biggest cause of lung cancer, after smoking, according to the World Health Organization. Smoke deposits in your lungs radioactive atoms of polonium-210 and lead-210, which are present in tobacco leaves.

Even your body contains small concentrations of radioactive atoms, including potassium-40 in soft tissue and lead-210 in your skeleton. You continuously absorb radioactive atoms from food, including vegetables fertilized with phosphate. You are what you eat: therefore you are radioactive.

You are also continuously penetrated by particles and radiation, mainly produced by cosmic rays originating from the radioactivity of the Sun, or from the outer reaches of our galaxy. Although the atmosphere and the Earth's magnetic field are shielding your body from this cosmic shower, you are still exposed to weak secondary cosmic radiation, which is more intense if you are at high altitudes, where the atmosphere is much thinner. Every second, at least one muon, a particle like an electron but about 200 times heavier, crosses your body. If radioactivity produced sound, the noise would be unbearable. Listening to the 'sounds' of radioactivity requires special instruments developed by nuclear and particle physicists. The history of radioactivity in the past century or so is linked to the progress of radiation and particle detectors, from photographic plates and electrometers, used by the pioneers of radioactivity studies in the 19th century, to modern semiconductors and scintillators, interfaced to computers by advanced microelectronic circuitry.

Many materials we touch have a small amount of radioactivity, but most of the time the effect is so small that even sophisticated detectors cannot measure it.

One group of scientists is particularly concerned about radioactivity and radiation. They are the physicists in search of particles and decays so rare that they can be detected only in a place where there is a very low background radioactivity and radiation. One of the most 'silent' places in the world is the underground laboratory of the Italian Institute of Nuclear Physics, built under the Gran Sasso Mountain in central Italy. If you take a detour from the A24 highway, going from Rome to L'Aquila, the tunnel through the mountain takes you to three cathedral-sized halls, each 100 metres long and almost 20 metres high. They are packed with sophisticated particle detectors like those used at CERN's Large Hadron Collider. A barrier of rock, 1,400 metres thick, reduces the flux of cosmic radiation a million times. In addition, the dolomite rocks of the mountain have natural radioactivity levels, from uranium and thorium, thousands of times lower than those on the Earth's surface.

The Italian Gran Sasso Laboratory can perform very challenging experiments. One detector, Opera, can pick up the elusive neutrinos (low-mass, electrically neutral fundamental particles) shot against the Gran Sasso Mountain from the accelerator located at the CERN European laboratories 732 kilometres away. In 2011, Opera's physicists announced that they observed CERN's neutrinos travelling faster than the speed of light, breaking a barrier established by Einstein's law of special relativity. This result is presently disputed after the discovery of anomalies in the measurement procedure.

Special materials are selected to enhance the sensitivity of the Gran Sasso detectors. For example, lead from 120 ancient Roman ingots is used in the detector CUORE (Cryogenic Underground

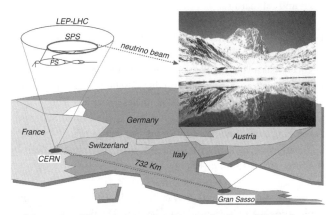

1. Schematics of the system used to measure the time of flight of neutrinos from CERN (Geneva) to the Gran Sasso Laboratories (Italy) over a distance of 732 kilometres

Observatory for Rare Events) in a study to measure the mass of the neutrino. The ingots were recovered from ships that sank 2,000 years ago off the coast of Sardinia. Weighing about 33 kilograms each, they were originally destined to become coins, water pipes, or projectiles for catapults. Freshly mined lead is slightly radioactive, as it contains lead-210, part of the uranium natural decay series, with a half-life (the time for the number of radioactive nuclei to halve) of 22.3 years. However, the original radioactivity of the Roman ingots had disappeared after two millennia of radioactive decay.

Recently, one of the Gran Sasso detectors intercepted, for the first time, a few antineutrinos produced by the radioactivity of uranium in the core of our planet, thousands of kilometres under our feet. These so-called geo-neutrinos bring precious new information about the interior of the Earth. The sensitive neutrino detector BOREXINO can also see man-made neutrinos originating from the 435 nuclear power reactors in operation around the planet and from hundreds of small reactors used for research, radiopharmaceuticals production, and for powering ships and submarines.

Detectors like BOREXINO could be used as global nuclear monitors, a sort of 'Big Brother' peering into nuclear reactor fuel to check that it is not used for illegal activities. Each year, the world's nuclear power reactors produce about 20,000 kilograms of plutonium, a primary material in nuclear weapons. The Lawrence Livermore National Laboratory in the United States is developing a neutrino detector for nuclear safeguards applications, to be used by the watchdogs at the United Nations, the International Atomic Energy Agency.

There is another 'Big Brother' watching global radioactivity – the International Monitoring System, developed by the Comprehensive Test Ban Treaty Organization in Vienna, to make sure that countries don't test new nuclear bombs. It consists of a global network using radioactivity detection, along with seismology, hydroacoustics, and infrasound measurements, to pick up any sign of a nuclear explosion. Eighty stations worldwide are being set up to monitor the possible presence of radioactive particles in the atmosphere.

Even peaceful nuclear infrastructure can arouse global fear. In March 2011, millions of people watched in horror the images of the Fukushima nuclear power plant disaster after a magnitude 9 earthquake and tsunami slammed Japan. They saw plumes of fallout, including caesium-137 and iodine-131, being transported thousands of kilometres away from the stricken reactors. The CTBTO monitoring system in Vienna gave early warning of the radioactivity ejected from the nuclear power plant, but to no avail.

2. The mushroom cloud above Bikini Atoll on 25 July 1946, following an atomic bomb test. Some ships were placed near ground zero to test the effects of the nuclear explosion

Nuclear radioactivity has many sides: some bright, others dark. As we shall see, it can help scientists to piece together the evolution of the universe, of the Earth, and of our own species. It can generate energy, boost food security, and enhance health. Most hospitals have a nuclear medicine department for cancer diagnostics and therapy. On the other hand, radioactivity could also annihilate mankind.

This is a brief story of our engagement with the awesome power of the atomic nucleus.

Radioactivity: A Very Short Introduction

Chapter 1

Opening the nuclear Pandora's box

You must take some precautions when consulting Marie Skłodowska Curie's laboratory notebooks in the French National Library. They are contaminated with radioactivity of several becquerel/cm^2, mainly where the great scientist held them. You have a high risk of receiving a radiation dose above the values recommended by the ICRP. She enjoyed cooking at home, so her cookbooks are also highly radioactive. During the last years of her life, she was attending international conferences with bandages on her hands, which were badly injured with radiation burns. Marie ingested and breathed large amounts of pitchblende dust in her laboratory, and this had probably caused the pernicious anaemia that killed her in 1934. Her daughter, Irène, who worked in the same laboratory, also died relatively young of leukaemia, as a result of overexposure to radiation.

Marie Curie's hands were probably heavily contaminated on 6 February 1898 when she angrily wrote in her notebook, with ten exclamation marks, that the temperature in the laboratory was 6.25°C.

In December 1897, she had started her thesis, working with her husband Pierre Curie. Marie was enthusiastic about the project, even in the miserable conditions of her small laboratory at the

École Supérieure de Physique et de Chimie Industrielles de la Ville de Paris. She was keen to understand the unexpected discovery made by Henri Becquerel, professor of physics applied to natural sciences in the Museum d'Histoire Naturelle. Two years earlier, Becquerel had noticed that uranium minerals had blackened Lumière photographic plates, wrapped in black paper, even without prior exposure to sunlight. This implied the existence of a new form of radiation that the scientist called 'uranic rays'.

Becquerel had originally meant to study the properties of fluorescent substances, like uranium salts, searching for radiation like the X-rays discovered by Wilhelm Conrad Röntgen in 1895 at the University of Würzburg in Germany. Röntgen, who was studying the effects of electric discharges through gas at low pressure, observed that a fluorescent screen (a sheet of paper coated with barium platinum-cyanide) was glowing, during the discharge, even when it was shielded or located in the next room. Becquerel was convinced that similar mysterious rays might be emitted from naturally fluorescent substances. This line of research was preoccupying many scientists during the second half of the 19th century. They were experimenting on the so-called 'cathode rays' observed in discharge tubes, where a vacuum was created by primitive mercury pumps, while a high voltage was produced by transformers based on Rühmkorff coils. At that time, fluorescent materials, photographic plates, and electroscopes were the only radiation detectors available.

In 1897, the British physicist Joseph John Thomson demonstrated that cathode rays were negatively charged particles. Deflecting them with electric and magnetic fields, he measured the ratio of their electric charge to their mass. He thus discovered the first subatomic particle, the electron, which has a mass 1,837 times smaller than that of the hydrogen atom. For this discovery, he was awarded the Nobel Prize in 1906.

Soon, X-rays overshadowed cathode rays, exciting both the scientific community and the public. The Kaiser Wilhelm II invited Röntgen to court to demonstrate his 'new kind of rays', and decorated him with the Prussian Order of the Crown. The image of the bony hand of Anna-Bertha Röntgen became an icon, reported in many textbooks throughout the past 110 years. 'I have seen my own death' was her famous exclamation on seeing the first radiographic image of a human body part, complete with wedding ring.

Becquerel did not find what he was looking for, but he discovered the spontaneous radioactivity of uranium. The subsequent work of Marie and Pierre Curie revealed that other natural elements had this special property.

Becquerel and his colleagues also noticed that radioactivity caused the discharge of electroscopes, that is, instruments measuring electric charges through the effects induced by the electrostatic force (that can move, for example, two charged gold leaves, displacing them by an amount proportional to their charge). Back in 1785, the French scientist Charles-Augustin de Coulomb had already discovered that an electroscope could spontaneously discharge. Just over one century later, this phenomenon would be finally explained as the effect of ionising radiation coming from the Earth (but also from the cosmos, as we shall see later).

The word 'radioactive' appears for the first time in the paper, 'On a New Substance, Radioactive, Contained in Pitchblende', published by Pierre and Marie Curie in July 1898.

The pitchblende – from the German *Pechblende*, ' bad-luck rock' – was coming from Sankt Joachimsthal, Bohemia, now Jáchymov in the Czech Republic, where, in 1789, a Berliner pharmacist, Martin Klaproth, had isolated a new element that he named uranium, after the newly discovered planet Uranus. At Joachimsthal, mining goes back to 1516, when rich veins of silver were discovered. Many

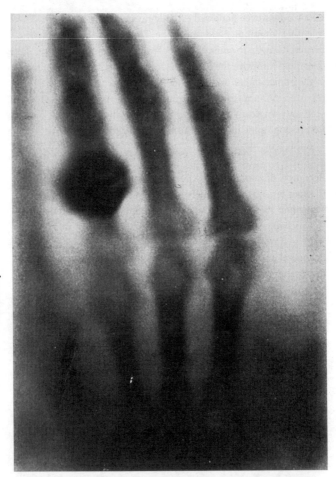

3. The X-ray radiography of Anna-Bertha Röntgen's hand, taken by
Wilhelm Röntgen on 22 December 1895

miners were affected by a strange disease, attributed to 'foetid
vapours' and vicious 'subterranean dwarfs'. In 1879, the disease
was possibly diagnosed as 'malignant cancer' to the lungs; but it
wasn't until 1932 that the *American Journal of Cancer* confirmed

that 'the most probable cause of the tumours, radium emanation, which is contained in the air of Jáchymov pits up to 50 mache units'. (The mache unit is an old unit for volume radioactivity, equal to 13.45 becquerel per litre.)

The new substance, isolated by the Curies, would be called polonium, in honour of Marie's country of origin. As a matter of fact, Poland did not exist at the time, as it had been divided between Prussia, Russia, and the Austrian Empire. For the first time in the history of science, a new, invisible, element was identified only by the rays it emitted. Polonium-210, a decay product of uranium-238, is the most abundant isotope of polonium.

The success of their experiments was due to the use of methods that were more quantitative than Becquerel's photographic plates. The Curies were using electrometers to measure the ionization of air induced by radioactive substances, which was proportional to the intensity of the 'uranic rays'. Very small radioactivity amounts could be quantified using the 'piezoelectricity' that Pierre Curie and his brother Jacques had discovered in 1880.

The unit for radioactivity is named after Henri Becquerel. One becquerel (Bq) equals one radioactivity decay (or nuclear disintegration) per second. This unit is used in the Standard International System, replacing the old unit, the curie (Ci), equivalent to 37 billion Bq, approximately the activity of one gram of radium.

In a paper published in December 1898, 'On a New, Strongly Radioactive, Substance Contained in Pitchblende', Pierre and Marie announced the discovery of radium (which means 'ray' in Latin). Now we know that radium-226, the main isotope of radium, is a decay product of uranium-238 and decays into radon-222. Initially, the new element was not available in visible amounts. Four years later, Marie would extract one-tenth of a gram of radium from a ton of pitchblende. In the following years, radium would become a very popular element. Many thought that the substance, which emitted invisible rays and glowed an eerie blue in the dark, would be a panacea to add to food and drink.

The Curies won the Nobel Prize in 1903, 'in recognition of the extraordinary services they have rendered by their joint research on the radiation phenomena discovered by Professor Henri Becquerel'. In January 2011, during the UNESCO launch of the International Year of Chemistry, Hélène Langevin-Joliot, Marie's granddaughter, gave a speech relating how, after Pierre's and Marie's discovery, the French Academy of Science had transmitted only the names of Pierre Curie and Henri Becquerel to the Nobel Committee. Eventually, after the energetic protest of her husband, Marie was finally included in the nomination.

It didn't take long, though, for the pioneers of radioactivity to realize how dangerous this phenomenon was. During the Nobel Prize ceremony, Pierre Curie compared his discovery to Nobel's own invention of dynamite. 'It can even be thought that radium could become very dangerous in criminal hands, and here the question can be raised whether mankind benefits from knowing the secrets of Nature, whether it is ready to profit from it or whether this knowledge will not be harmful for it', he said. More than a century later, the question of our maturity to handle nuclear matters remains unanswered.

In 1911, five years after Pierre's accidental death, Marie Curie received her second Nobel Prize, this time in Chemistry, 'in recognition of her services to the advancement of chemistry by the discovery of the elements radium and polonium, by the isolation of radium and the study of the nature and compounds of this remarkable element'. By the beginning of the First World War, Marie had stored most of the world's supply of radium, which she took to a bank in Bordeaux.

In the same year, 1898, when the Curies announced their radioactivity discoveries, Ernest Rutherford, a New Zealand-born scientist working as research student at the Cavendish Laboratory in the UK, produced other exciting news. He had observed two kinds of rays emitted by uranium: alpha rays, which could be easily stopped, and beta rays, much more penetrating.

To verify the absorption properties of the different radiations, Rutherford covered the uranium material with aluminium sheets of different thickness, using an electrometer to measure the electric current produced by the transmitted radiation. Becquerel was convinced that uranium radiation had different components, but he could not identify them. In 1899 and 1900, he went back to the study of the beta particles emitted by uranium, deflecting them in magnetic fields. Their deflection depended on their mass and charge, following basic laws of classical physics known at the time. With these experiments, he showed that beta particles had the same charge to mass value as Thomson's electron.

In 1900, the French physicist Paul Villard discovered more penetrating rays. Rutherford showed later that they were electromagnetic radiation similar to X-rays but with shorter wavelength, and called them gamma rays.

In 1908, Rutherford demonstrated that alpha rays were energetic helium atoms stripped of all their electrons. Radium, a strong alpha-emitter, was placed in a tube with very thin walls, sealed inside a larger container with thicker walls. The alpha particles could escape from the thin-walled tube, but remained trapped in the larger tube. A glow discharge could show the atomic spectra typical of helium, which was not initially present.

In the same year, Rutherford received the Nobel Prize in Chemistry 'for his investigations into the disintegration of the elements, and the chemistry of radioactive substances'.

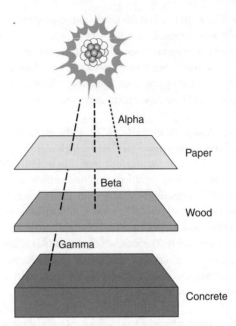

4. Examples of the penetrating power, in different materials, of alpha particles, beta particles, and gamma rays

During the early days of her discovery, Marie Curie did not know that radioactivity was due to a transmutation made possible by the nuclear structure of the atom. She had to wait for further investigations involving again the scientist from New Zealand.

The nucleus

It is as if you fired a 15-inch shell at a piece of tissue paper and it came back and hit you.

Ernest Rutherford could not believe that the alpha particles bombarding a very thin gold foil bounced back. So far, he had believed in the model proposed by his own professor, J. J. Thomson: the atom was made of electrons, with a negative charge, and some kind of positive charge uniformly filling a sphere one-ten-millionth of a millimetre across; this was the famous 'plum pudding' model. This model could not explain the results of the recent experiment, though, as alpha projectiles should have undergone small deflections as they crossed the pudding.

The discovery came utterly unexpected, during experiments performed in 1909 by his collaborator, Hans Geiger, and a research student, Ernest Marsden, at the University of Manchester. They were using a small chamber, under vacuum, containing an ampoule of radon-222 as a source of alpha particles, a gold foil as target, and a particle detector. The latter consisted of a glass screen covered with zinc sulphide. A few years earlier, while experimenting in his home laboratory, a gentleman scientist, Sir William Crookes, had discovered that radiation emitted by radon made the zinc sulphide glow. The effects were so beautiful that he invented a radioactive kaleidoscope, called a 'sphintariscope', which was sold in London shops. He assumed that beta and gamma rays produced the weak uniform glow, while alpha particles were causing the individual scintillations. Marsden and Geiger had adopted this technique for their experiment.

The microscope used to amplify the image of the small sparks could rotate around the cylindrical box to count the number of alpha particles emitted by the gold foil at different angles, when they struck the scintillator. The scientists had to work in a dark laboratory, where it took about 30 minutes for their eyes to adjust enough to see the small traces. Students, and sometimes women from outside the academic circle, were involved in this tedious work. Later, Geiger invented his Geiger counter, measuring the charge produced in a gas by the ionizing radiation. That made life much easier for the experimenters.

On Rutherford's advice, Marsden was checking for the possibility that alpha particles could scatter at large angles. After a few days in the dark laboratory, he was able to report the exciting news that

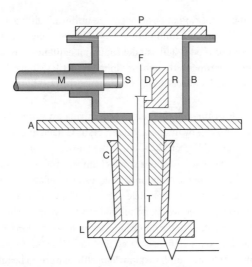

5. Geiger and Marsden's apparatus, used by Rutherford to discover the nucleus. It consists of a small chamber, under vacuum, containing an ampoule of radon-222 as source of alpha particles (R), a gold foil as target (F), and a scintillator (S) attached to a microscope (M)

some alpha particles were deflected by more than 90 degrees after a single encounter. These large deflections could be explained only by assuming that the positive charge in the gold atom was not distributed uniformly on a sphere the size of the atom, but concentrated instead in a small sphere, having a mass larger than that of the alpha particles. Earlier experiments with beta particles, carried out by a collaborator of J. J. Thomson, had been used to support the 'plum pudding' model, assuming that the final deflection was the result of multitude scattering by the atoms in the material.

In 1911, Rutherford proposed a new model for the atom, heralding the beginning of nuclear physics. According to this model, the atom has a nucleus with a diameter 10,000 times smaller than that of the atom, that is, about 14 femtometres

(1 fm = 10^{-15} m) for gold. The nucleus also includes most of the atomic mass and all the positive charge. He did not use the word 'nucleus', though. Instead, he stated that 'the atom consists of a central charge supposed concentrated at a point'. The much lighter electrons were distributed over the atom's volume outside the nucleus. The alpha particles could therefore cross the empty space between nuclei without being deflected; they were scattered at large angles only when they approached the central field of the nucleus. Four years later, Niels Bohr added a new feature to Rutherford's model of the atom: the 'quantization' of the orbits of the electrons, which confined the particles to certain orbits. This assumption was necessary to explain the stability of atoms; indeed, it was one of the paradigm shifts that marked the beginning of quantum physics.

The next step was to understand the structure of the nucleus and to control its transmutations.

Nuclear transmutation

With what we know today of the structure of atoms, we understand perfectly the hopeless task undertaken by alchemists, striving to transmute the different elements one to another, and to transform lead and mercury into gold. With the means at their command, they could not work on the essential part of the atom, that is to say the nucleus.

This was the incipit of Professor H. Pleijel, Chairman of the Nobel Committee for Physics of the Royal Swedish Academy of Sciences, on 10 December 1938. He made these comments in his prize award ceremony speech, before His Majesty King Gustavus Adolphus awarded the Nobel Prize to Enrico Fermi.

Thales from Miletus, 7th century BC, was the first known philosopher who tried to explain, without calling on some gods, the structure and behaviour of matter. Others continued on this line of thinking, including Leucippus and Democritus (5th century BC). The latter is remembered for his famous quote 'Nothing exists except atoms and empty space; everything else is opinion'. By Aristotle's time, in the 4th century BC, Greek philosophers had reached the conclusion that matter was made from the combination of four elements: earth, water, fire, and air. These elements could 'transmute', or change, under the action of cold, heat, and other environmental conditions. This is the genesis of alchemy, which was developed first by Greek, and then by Arab, scholars, with a silent interlude during both the pagan and the Christian phases of the Roman Empire.

The transmutation of one substance into another was called *khymeia*, in Greek. An active practitioner of the art of *al khymeia* (later called *alchemia* in Latin) was the Persian Jābir ibn Hayyān, who lived in the 8th and early 9th centuries. One of the theories of Jābir, also known as Geber, assumed that all metals were composed of mercury and sulphur.

Alchemy became popular again among the European scholars of the 13th century, when it was largely the preserve of clergy, who could comprehend the Arabic texts. Roger Bacon, a Franciscan from Oxford, was one of the best-known alchemists of the period. Supported by the Church, the alchemists were very active experimenters, investigating how matter behaved, but also trying to generate gold out of more ordinary materials. During the following centuries, alchemists continued their work. Some were serious natural philosophers, attempting to understand matter and the universe. Others were crooks and fakers promising philosopher's stones and miraculous elixirs. A well-known alchemist was Philippus Aureolus Paracelsus (1493–1541), a pioneer in the use of minerals and other substances in medicine.

The Irishman Robert Boyle (1627–91) played a key role in the evolution of alchemy. He wanted to separate alchemy from the 'magic' practices that had given it a bad name. He used the name 'chemistry' for the new discipline, dropping the Arabic article. Boyle proposed practical methodologies to identify the elements of matter, defining them as substances that could not be separated into simpler substances. The new natural philosophers, who were increasingly taking on the attributes of modern scientists, were able to demonstrate that earth, water, air, and fire were not elements according to Boyle's definition. Immediately, they could demonstrate that the Earth was made of tens of basic elements. During the 18th century, the French nobleman Antoine Lavoisier was able to show that air was a mixture of oxygen and nitrogen. Lavoisier was also convinced that water was made of hydrogen and oxygen, but he was guillotined during the French Revolution before this could be demonstrated.

In 1803, the English scientist John Dalton proposed to consider the atom as the basic constituent of matter, in which each element was characterized by a different atom. Atoms appeared again as the basic components of matter, more than 2,300 years after Greek philosophers had conceived them using only the laws of logic. Dalton wanted to go further, developing experiments that could identify the behaviour and properties of the atoms.

The next step was to give some order to the atoms, according to the elements they were representing. This was done in 1869 by the Russian chemist Dmitri Mendeleev, who arranged all known elements in his famous periodic table, on the basis of their specific properties.

The atoms, their components, and their capacity of transmuting could finally be observed at the onset of the 20th century, thanks to the discovery of radioactivity.

A plaque at McGill University in Montreal states:

> At this location, Ernest Rutherford and Frederick Soddy, during
> 1901–03, correctly explained radioactivity as emission of particles

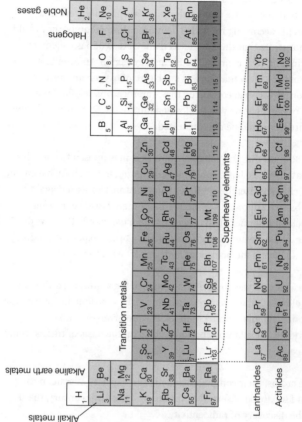

6. The periodic table of chemical elements, first developed by Mendeleev in 1869

from the nucleus and established the laws of the spontaneous transmutation of the elements.

Soddy had graduated in chemistry at Oxford before going to Montreal and would be awarded the Nobel Prize in Chemistry in 1921 'for his contributions to our knowledge of the chemistry of radioactive substances, and his investigations into the origin and nature of isotopes'. According to Soddy, isotopes were 'elements with identical external electronic systems, with identical net positive charge on the nucleus, but with nuclei in which the total number of positive and negative charges and therefore the mass is different'. Eleven years later, it would be shown that different isotopes have the same number of protons, but different numbers of neutrons.

Considering the facts accumulated since the discovery of radioactivity, Rutherford and Soddy concluded that the atoms of the radioactive elements, unlike those of the ordinary elements constituting the materials around us, were unstable, disintegrating with the emission of alpha or beta particles from the nucleus. The new atom left after the radioactive decay had physical and chemical properties different from the initial parent atom.

For example, the unstable uranium nuclide of mass 238 may decay and expel an alpha particle, a helium nucleus of mass 4, leaving a lighter atom of a different element, thorium, of atomic mass 234. This also disintegrates, emitting a beta particle and generating protactinium, an element of the same mass. Radioactive transitions finally transform the original atom of uranium into a stable atom of lead.

During these transformations, nuclei could be left in an excited state until they finally release their extra energy with the emission of gamma rays.

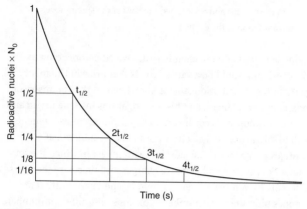

7. The law of radioactive decay, formulated by Rutherford and Soddy in 1902

The two scientists also discovered the general law of radioactivity:

$$N(t) = N_o e^{-\lambda t}$$

where the number of radionuclides at time t, N(t), is given in terms of N_o, the number of radionuclides at t = 0, and the characteristic decay constant λ.

This law can be obtained assuming that radioactivity is a statistical process. If we consider an ensemble of N(t) radionuclides at time t and we assume that each radionuclide has the same probability λ of decaying per unit time, the number dN of radionuclides decaying in time dt is:

$$dN = -\lambda N(t)dt$$

Integrating this equation one obtains the exponential decay law of radioactivity found experimentally by Rutherford and Soddy.

The decay constant λ is related to the half-life $T_{1/2}$ by:

$$T_{1/2} = \ln 2 \, / \, \lambda$$

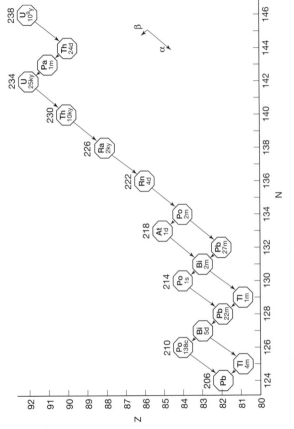

8. **Decay chain of uranium-238**

So far, we have been discussing the transmutation of unstable natural substances. It was again Lord Rutherford who, in 1919, demonstrated for the first time the artificial transmutation of elements. He was able to transform the stable element nitrogen into oxygen, finally realizing the alchemist dream. This was done by bombarding nitrogen with energetic alpha particles from radium. He noticed that occasionally an alpha particle got stuck to the nucleus while a fast proton was ejected, leaving a nucleus of oxygen-17. The rare form of heavy oxygen created by Rutherford is very scarce in nature (0.0373%); it was measured for the first time in the atmosphere in 1929.

The nuclear reaction performed by Rutherford can be represented with the following equation:

$$\,^{4}_{2}He_{2} + \,^{14}_{7}N_{7} \;\rightarrow\; \,^{17}_{8}O_{9} \;+\; \,^{1}_{1}H_{0}$$

where the number on the upper left-hand side of the symbol of the element corresponds to the total number of neutrons and protons (the 'mass number'), the number on the lower left-hand side corresponds to the number of protons in the nucleus (the 'atomic number'), and the number on the lower right-hand side corresponds to the number of neutrons, which are neutral particles in the nucleus. Neutrons would be discovered a few years after Rutherford's transmutation experiment, as we shall see.

In the same year, 1919, Jean Perrin suggested that nuclear reactions transmuting hydrogen into heavier elements might be responsible for the production of energy in the Sun and in other stars. It would take several decades to follow this idea and to show that nuclear transmutations are the basic mechanisms in the evolution of matter in the universe.

The first nuclear reaction was followed by many others, including the reactions that occurred in the laboratory of Irène Curie and her husband Frédéric Joliot. In 1932, they noticed that the bombardment of boron with alpha particles from a polonium source was producing positrons, e^+, and neutrons, n.

Positrons were the electron's anti-particles discovered in the same year by Carl Anderson, at the Californian Institute of Technology; neutrons were the neutral particles with the same mass as the proton, discovered by the British physicist James Chadwick some months before. Chadwick found the neutron while analysing the results of previous experiments by the Joliot-Curies, when they bombarded beryllium with alpha particles. The Joliot-Curies had wrongly interpreted the penetrating particles produced as being high-energy gamma rays rather than neutrons. The physicist Walther Bothe in Berlin had obtained similar results two years earlier. Chadwick carefully repeated these experiments using alpha particles from polonium and concluded that the reaction produced by Bothe and the Joliot-Curies was:

$$ {}_2^4 He + {}_4^9 Be_5 \rightarrow {}_6^{12} C_6 + {}_0^1 n_1 $$

The husband-and-wife team had more success with their nuclear reactions on boron. They identified two, simultaneous nuclear processes:

$$ {}_2^4 He + {}_5^{10} B_5 \rightarrow {}_7^{13} N_6 + {}_0^1 n_1 $$

$$ {}_7^{13} N_8 \rightarrow {}_6^{13} C_7 + e^+ $$

This experiment led to an exciting discovery: a *new* form of radioactivity.

Artificial radioactivity

Irène and Frédéric noticed that the emission of positrons persisted after the removal of the radioactive polonium source. The nuclear reactions between alpha particles and boron had produced nitrogen-13, an unstable nuclide that does not exist in nature but which can be created artificially, and chemically separated. They dubbed this new form of nitrogen, which transmutes with a half-life of 9.97 minutes into carbon-13 with the emission of a positron, radionitrogen. This was the discovery of 'artificial radioactivity'. Using the same method, the Joliot-Curies bombarded aluminium and magnesium with alpha sources, producing radioactive isotopes of phosphorus (radiophosphorus or phosphorus-30, $T_{1/2}$ = 2.50 minutes) and silicon (radiosilicon or silicon-27, $T_{1/2}$ = 4.16 seconds), respectively. In these cases, they could separate chemically the radioactive substance from the unchanged atoms constituting the bulk of the bombarded material.

In February 1934, they published their work in *Nature*, receiving the Nobel Prize in December 1935 'in recognition of their synthesis of new radioactive elements'. In that same paper, they suggested that these and other radioactive substances could be produced with other bombarding particles, such as protons, deuterons, and neutrons.

Infact, while Joliot and Curie were experimenting with alpha particles from natural radioactive substances in nuclear reaction studies, at Berkeley's cyclotron laboratory, Ernest Lawrence, Stanley Livingston, and other American physicists were doing similar experiments with high-energy deuterons (nuclides of isotopic mass 2 and atomic number 1, sometimes called heavy protons), observing the production of artificial radioactivity. A flurry of similar experiments developed independently in many other laboratories, using radioactive sources and the newly invented ion accelerators.

The accelerators used for producing high-energy beams of charged particles were based on different principles. The first class of systems was based on the use of high voltage, with the ions accelerated through a large potential difference in a vacuum tube. Cockcroft-Walton systems (based on voltage multipliers), built at the Cavendish Laboratories, could reach several hundred thousand volts, while Van de Graaff generators (accumulating high voltages with moving belts), built at Princeton University, could produce voltages above a million volts. In the second class of accelerators, the ions were accelerated along an electromagnetic wave. A third class used multiple acceleration obtained with a resonance of an electric field. Berkeley had the most advanced systems of this kind, based on the use of large electromagnets. They could accelerate deuterons to energies of several million electron volts.

An exciting line of work was being developed in Rome, in the famous physics laboratory of via Panisperna (the building is presently part of the Italian Ministry of the Interior), where Enrico Fermi and the other 'ragazzi' had learned how to evaporate polonium on beryllium foils to make neutron sources via the reaction that we mentioned earlier (page 19)

$$_2^4 He + {}_4^9 Be_5 \rightarrow {}_6^{12} C_6 + {}_0^1 n_1$$

In the first months of 1934, Enrico Fermi (or 'the Pope', as he was nicknamed in the group to acknowledge his scientific authority) used neutrons to produce artificial radioactivity. The neutron flux was much lower than that of alpha particles, and although the reaction had a much larger cross section, the scientists were unable to see any results. The cross section is a measure of the probability of occurrence of a nuclear reaction and is measured in barns – with 1 barn = 10^{-28} m^2. Fermi replaced the polonium-beryllium source with a more powerful radon-beryllium source, built by filling a glass bulb with radon and beryllium powder. As a result, the team was finally able to produce several radioactive substances.

The scientists irradiated elements of increasing atomic number, up to thorium and uranium, believing they would produce new transuranic elements. In fact, at the beginning of summer 1934, they went on holiday convinced they had produced a new element,

the so-called 'element 93'. They were unaware they had produced
reactions of nuclear fission – the splitting of the uranium nucleus
into smaller nuclei with the emission of particles and release of
energy – even though they had been alerted to this possibility by
their German colleague Ida Noddack, a distinguished chemist and
physicist who would be nominated several times for the Nobel
Prize.

Fermi won the Nobel Prize in 1938 for 'his demonstrations of the
existence of new radioactive elements produced by neutron
irradiation, and for his related discovery of nuclear reactions
brought about by slow neutrons'.

In December of the same year, it was the German scientist Lisa
Meitner and her nephew Otto Robert Frisch who came up with
the idea that a uranium atom would split into two lighter
fragments, after the absorption of a neutron. The splitting of the
atom could explain the production of barium isotopes in reactions
between neutrons and uranium, which had just been discovered
in the experiments performed by their colleagues, Otto Hahn and
Fritz Strassmann, in Berlin. According to Meitner and Frisch, the
capture of a neutron in uranium created a large 'compound
nucleus' with all the neutrons and protons excited by the energy
deposited by the additional neutron. The compound nucleus was a
concept just developed by Bohr, with whom Frisch was working in
Copenhagen. Uranium would elongate and 'fission' into two
fragments, like a drop of water. Hahn would be awarded the 1944
Nobel Prize in Chemistry 'for his discovery of the fission of heavy
atomic nuclei'. Bohr nominated Meitner and Frisch for the Nobel
Prize in 1946, without success.

Meitner and colleagues published their results in *Nature* in 1939,
the same year Hitler invaded Poland, starting the Second World
War. In the same year, Joliot and other scientists published in
Nature that 'recent experiments have shown that neutrons are
liberated in the nuclear fission of uranium induced by slow

neutron bombardment', the effect that was needed to produce a chain nuclear reaction. The possibility of making available a weapon of mass destruction to Hitler was on the minds of scientists, on both sides of the conflict.

After receiving his Nobel Prize in Stockholm, Fermi migrated to the United States and started working on the construction of a nuclear reactor. This would have been a key activity in the Manhattan Project, which started in 1942, with the aim of constructing the first nuclear bomb.

The nuclear 'Pandora's box' was now fully open. An increasing variety of radioactive nuclides was produced and added to the radiation naturally existing in the environment.

The radioactive environment

The main naturally occurring radionuclides of primordial origin are uranium-235, uranium-238, thorium-232, their decay products, and potassium-40. The average abundance of uranium, thorium, and potassium in the terrestrial crust is 2.6 parts per million, 10 parts per million, and 1% respectively.

Uranium and thorium produce other radionuclides via neutron- and alpha-induced reactions, particularly deeply underground, where uranium and thorium have a high concentration. Long-lived radionuclides produced by these reactions, including beryllium-10, carbon-14, chlorine-36, and aluminium-26, have been observed in uranite and pitchblende with ultra-sensitive atom-counting methods, which will be discussed in the following chapters. The radionuclide iodine-129, produced by the spontaneous fission of uranium-238 and neutron-induced fission of uranium-235, can also be measured in uranium minerals.

A weak source of natural radioactivity derives from nuclear reactions of primary and secondary cosmic rays with the atmosphere and the lithosphere, respectively. They include carbon-14, beryllium-10, and other long-lived geo-chronometers which we will discuss in Chapters 3, 7, and 8. Accretion of extraterrestrial material, intensively exposed to cosmic rays in space, represents a minute contribution to the total inventory of radionuclides in the terrestrial environment.

Nuclear weapon tests in the atmosphere and underground have also introduced a large variety of radionuclides into the environment. Short-lived radionuclides have decayed away, but some of the long-lived radionuclides, such as carbon-14 and plutonium-239, are still around.

Other nuclear activities, including nuclear power reactor operation and decommissioning, fuel reprocessing, and nuclear waste disposal, are all sources of environmental radioactivity.

Accidents at nuclear power plants can also contribute to environmental radioactivity. The 1986 Chernobyl accident was the most severe event in the history of nuclear power, blasting a large amount of radionuclides across the northern hemisphere. The cloud from the burning reactor carried fission products, particularly iodine-131 ($T_{1/2}$ = 8 days) and caesium-137 ($T_{1/2}$ = 30.1 years), over much of Europe. The latter is still detectable in soils and certain foodstuffs. More than 5 million people still live in regions of Ukraine, Belarus, and Russia that have a radioactivity from caesium-137 of more than 37,000 becquerel/m². In north-eastern Italy, where rain dumped large amounts of caesium-137 and other radionuclides on the ground in May 1986, the radioactivity of mushrooms, wild berries, and game is still monitored.

Recently, the fear of nuclear reactors being hit by extreme environmental or geologic events has been mounting, particularly in the wake of the near meltdown, in March 2011, at the Fukushima power station, 200 kilometres from Tokyo, as a result of a massive earthquake and consequent tsunami. A large amount of radioactive material was dispersed into the environment. Caesium-137 reached 80,000 becquerel per kg in vegetables grown in the Fukushima prefecture, 160 times the legal limit. Iodine-131 was detected in Tokyo tap water at levels of 200 becquerel per litre, more than double the legal limit.

An increasing amount of radioactive nuclides, both natural and man-made, have been mobilized since the discovery of radioactivity. Radioactivity-based systems are of paramount socioeconomic importance in many sectors, including energy, health, industry, and agriculture. Strict measures are therefore needed to keep them safe and secure.

The international community, following guidelines determined by the IAEA, is heavily investing in improving the physical protection and accounting systems for radioactive materials. Neutron and gamma detectors are located at many border-control points. For example, the detection of neutrons would reveal the presence of uranium and other nuclear materials. Most radionuclides of interest emit gamma rays of relatively high energy that can be easily identified as they penetrate packaging material. Pure beta-emitters such as phosphorus-32, strontium-90, and yttrium-90 can be easily hidden during transport. In the event of discovery of radioactive material, nuclear forensics experts are mobilized to look for clues that might reveal its origins.

The protection of humankind and the environment from radioactivity has improved since Becquerel's and Curie's times. In the 1920s, the first radiation protection directives were issued in many countries. In 1928, the ICRP, just established in Stockholm, produced the first recommendations for radiation protection against 'injuries to the superficial tissues and derangements of internal organs and changes in the blood'. For several years, cumulative, long-term genetic effects were not considered in the development of protection policies.

A large body of knowledge is now available on radiation effects to the human body and to other organisms, particularly from the study of nuclear incidents and non-peaceful applications of radioactivity.

Hitting your body

Radiation can penetrate the tissues of your body, interacting with its atoms and molecules through mechanisms that depend on the nature of the bombarding rays. The effects of X-rays on the human body were recognized immediately after their discovery. First reports on X-ray burns appeared in the *British Medical Journal* in 1896. The effects of radioactivity became known soon after its discovery and were reported in scientific journals before the end of the century.

Charged particles such as protons, beta and alpha particles, or heavier ions that bombard human tissue dissipate their energy locally, interacting with the atoms via the electromagnetic force. This interaction ejects electrons from the atoms, creating a track of electron–ion pairs, or ionization track. The energy that ions lose per unit path, as they move through matter, increases with the square of their charge and decreases linearly with their energy, as was shown by Hans Bethe in 1930.

The energy of the particles considered here, either emitted from naturally occurring radioactive isotopes or produced by accelerators, is often measured in millions of electron volts (mega-electron-volts, or MeV). This is the energy that a particle would acquire when accelerated by a one million volt generator. An electron with the energy of 1 MeV has 94% the speed of light. An alpha particle or heavier ions with the same energy would have a much lower velocity.

An alpha particle from americium-241, with energy of 5.5 MeV, will penetrate the tissues of your body, producing tens of thousands of electron–ion pairs. These alpha particles would not be able to cross through your clothes and, if they hit your skin directly, they would stop in the outer layer, the epidermis, with a penetration depth of less than 50 microns.

Electron–ion pair formation has been used for electrometers and ionization chambers to detect radiation since the early days of radioactivity research. The formation of one electron–ion pair in the air requires 35 eV. An alpha particle with 5 MeV produces about 140,000 electron–ion pairs before stopping. More free electrons are created with other materials such as semiconductors that need only 2 to 3 eV to produce an electron–hole pair, making them very effective as radiation detectors, as even low-energy radiation would produce a measurable electric signal.

A high-energy X-ray or a gamma-ray photon can penetrate your body deeply. Sometimes the energy lost in a single event can knock an electron out of the atom. Loss of the full energy of the incoming photon produces the photoelectric effect. The loss of part of its energy produces the Compton effect. The resulting positive ion and the electron release their energy locally, interacting with the electric field of the atoms, as described above. When the photon energy is higher than 1.02 MeV, the nuclear field can convert the gamma ray into two massive particles. The electron will stop via the ionization process and the positron will annihilate, emitting two equal gamma rays in opposite directions.

Neutrons can also penetrate deeply and deposit their energy, interacting with a nucleus via the nuclear force. Contrary to the electromagnetic force, the nuclear force acts only at a very small range, but it is much stronger; this is why it is also called the 'strong' force. The encounter of neutrons with hydrogen nuclei, abundant in your body, which is mostly made of water, is particularly effective, with a large transfer of energy from the impinging neutron to the proton target. The resulting energetic proton slows down, like all charged particles, releasing energy locally along its path. Neutrons can also induce different nuclear reactions on the elements constituting your body, depending on their kinetic energy. For example, slow neutrons can stick to a hydrogen nucleus and produce a deuteron with a release of gamma rays.

Ionizing radiation can cause damage to your cells, disrupting chromosomes and breaking DNA chains. Gamma rays and neutrons penetrate your skin and reach all of your internal organs. Radioactive materials emitting this radiation are dangerous also at some distance. Radioactive sources emitting alpha particles become a serious health hazard only if you ingest them. The Russian KGB agent Alexander Litvinenko was murdered in 2006 with the ingestion of one microgram of polonium-210 ($T_{1/2}$ = 138 days) possibly given to him in a sushi restaurant in London. This radionuclide emits only alpha particles, which could be easily shielded from radiation monitors during international transport of the radioactive substance.

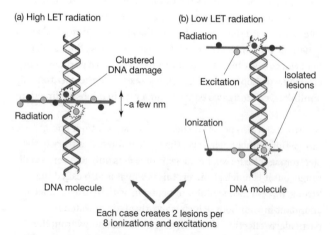

(a) High LET radiation

Clustered DNA damage

~a few nm

Radiation

DNA molecule

(b) Low LET radiation

Radiation

Isolated lesions

Excitation

Ionization

DNA molecule

Each case creates 2 lesions per 8 ionizations and excitations

9. DNA damage by ionizing radiation with low and high linear energy transfer

28

Your body has some natural radioactivity, including carbon-14 atoms, which emit beta particles that can damage some of the cells near the radioactive nuclei. Your carbon-14 radioactivity is 13.6 decays per minute per gram, which corresponds to 3,700 Bq, if your weight is about 70 kilograms. Other radionuclides present in your body are potassium-40 (4,000 Bq), uranium(2 Bq), polonium-210 (40 Bq), radium-226 (1.1 Bq), thorium (0.21 Bq), and tritium (23 Bq).

The energy deposited in the tissues and organs of your body by ionizing radiation is defined *absorbed dose* and is measured in *gray*. The dose of one gray corresponds to the energy of one joule deposited in one kilogram of tissue. The biological damage wrought by a given amount of energy deposited depends on the kind of ionizing radiation involved. The *equivalent dose*, measured in *sievert*, is the product of the dose and a factor w related to the effective damage induced into the living matter by the deposit of energy by specific rays or particles. For X-rays, gamma rays, and beta particles, a gray corresponds to a sievert; for neutrons, a dose of one gray corresponds to an equivalent dose of 5 to 20 sievert, and the factor w is equal to 5–20 (depending on the neutron energy). For protons and alpha particles, w is equal to 5 and 20, respectively. There is also another weighting factor taking into account the radiosensitivity of different organs and tissues of the body, to evaluate the so-called *effective dose*. Sometimes the dose is still quoted in *rem*, the old unit, with 100 rem corresponding to one sievert.

You can receive the dose of 1 sievert without feeling particularly ill. With a dose of 2 sievert to the whole body, you will feel nausea and your hair will fall out. With a dose of more than 2 sievert, you may die. With a dose of 3 sievert, your probability of dying is 50%. Doses below 1 sievert, which don't give immediate somatic effects, have long-term genetic consequences, though, increasing the probability of developing cancer. This is caused by changes induced in the genes that regulate cell multiplication.

The average dose you receive from natural radiation during one year is around 2,400 microsievert (one-millionth of a sievert) – 1,260 microsievert from radon, 480 microsievert from other environmental radioactivity, 390 from cosmic rays, and 290 from food. An ordinary radiography corresponds to 100 microsievert. The dose from an X-ray analysis at your dentist is 10 microsievert, the amount you get in a year from the natural radioactivity in your own body. You receive more radiation from an X-ray mammography, 1,000 to 2,000 microsievert, and from a CT scan, approximately 3,000 to 4,000 microsievert.

If you smoke 20 cigarettes per day, the additional annual dose from tobacco's radioactivity is 200 to 400 microsievert. Eating a banana will add 0.1 microsievert to your annual dose. You are not safe even in your bed, if you sleep with a partner, as you will receive an additional dose of 0.05 microsievert during the night. A return trip from Rome to Sydney will give you 500 microsievert. Reading a book produces a dose rate of about 0.01 microsievert per hour, mainly from potassium-40, much lower than the background radiation dose rate (0.1–0.4 microsievert per hour).

If you are professionally exposed to ionizing radiation, the ICRP recommends a dose limit of 20 millisievert (one-thousandth of a sievert) per year. The maximum annual permissible dose for a person in the general population is 1 millisievert. These figures are meant above the background, not considering medical exposure, for whole body exposure. These limits are much smaller than the first tolerance dose recommended by the ICRP in 1934, which was 500 millisievert for professionally exposed workers (only in 1949 was dose limitation for the public considered).

Professionally exposed workers can monitor very effectively the doses they are exposed to, caused by external sources of radiation. The most popular personal monitors are called thermo luminescent dosimeters, based on lithium fluoride crystals. Ionizing radiation causes some electrons in the crystal's atoms to jump to higher energy states, remaining captured in so-called electron 'traps', which have been created by adding specific impurities. Heating the crystal induces the electrons to fall back to their ground states, emitting light that is measured by a photomultiplier. These dosimeters are used to monitor gamma rays, neutrons, and beta particles. To provide the dose instantly, one can rely upon other personal dosimeters, like silicon diode detectors, based on the production of electron–hole pairs. They effectively replaced the old dosimeters based on the ionization of air. Finally, body contamination can also be checked, using whole-body or partial-body counting.

Chapter 2
Unlimited energy?

In 1903, Pierre Curie discovered that one gram of radium gives
out enough energy to boil one gram of water in about one hour.
The energy generated by pitchblende had been puzzling Becquerel
and Curie since the first years after the discovery of radioactivity.
It seemed that uranium and other naturally occurring radioactive
substances endlessly emitted energy, apparently violating
well-established thermodynamics principles.

It is now known that the laws of physics had, of course, not been
violated, and that the amount of energy generated by the natural
radioactivity of our planet is huge. Natural radioactivity is mainly
produced by uranium, thorium, and potassium. The total heat
content of the Earth, which derives from this radioactivity, is
12.6×10^{24} MJ (one megajoule = 1 million joules), with the crust's
heat content standing at 5.4×10^{21} MJ. For comparison, this is
significantly more than the 6.4×10^{13} MJ globally consumed for
electricity generation during 2011.

This energy is dissipated, either gradually or abruptly, towards the
external layers of the planet, but only a small fraction can be
utilized. The amount of energy available depends on the Earth's
geological dynamics, which regulates the transfer of heat to the
surface of our planet. The total power dissipated by the Earth is
42 TW (one TW – 1 trillion watts): 8 TW from the crust, 32.3 TW

from the mantle, 1.7 TW from the core. This amount of power is small compared to the 174,000 TW arriving to the Earth from the Sun.

The best scheme for geothermal power generation is based on the use of permeable fracture networks at several kilometres depth, with temperatures of 150–200°C and a large surface for heat exchange. These fractures can be artificially enlarged; injection wells are constructed to bring water from the surface, allowing heat exchange. Other wells are used to recover steam or hot water for electricity generation. These so-called 'enhanced geothermal systems' (EGS) could produce more than 100,000 MW of electricity in Europe. The single EGS power plant capacity can potentially be scaled up from the few MW of the existing systems to 100 MW.

It is difficult to evaluate the global potential of 'hot rock' energy for electricity production, but the technology will keep growing thanks to improvements in 3-D seismic and underground radar imaging to assess geothermal resources. New drilling technology based on lasers and high temperature flames will put deeper geothermal resources within reach.

About 40 countries in Africa, Central and South America, and the Pacific, all on the edges of tectonic plates and hence characterized by high volcanic and tectonic activity, could generate all of their electricity from geothermal energy.

The global geothermal electricity capacity installed has increased linearly in the past 40 years, approaching 11,000 MW in 2010. Recent scenarios forecast an electricity production of 1,400 TW-year by 2050.

Meanwhile, a man-made radioactivity process, nuclear fission, can multiply energy production by a large factor.

The dawn of nuclear energy

'The Italian navigator has landed in the New World.' This cryptic phrase, transmitted on 2 December 1942 to the US National Research Defense Committee, meant that Enrico Fermi and his team had won a secret race that had started at the University of Rome in 1934. They had achieved the first self-sustaining nuclear chain reaction to initiate the controlled release of nuclear energy.

Neutrons emitted during fission reactions have a relatively high velocity. When still in Rome, Fermi had discovered that fast neutrons needed to be slowed down to increase the probability of their reaction with uranium. The fission reaction occurs with uranium-235. Uranium-238, the most common isotope of the element, merely absorbs the slow neutrons. Neutrons slow down when they are scattered by nuclei with a similar mass. The process is analogous to the interaction between two billiard balls in a head-on collision, in which the incoming ball stops and transfers all its kinetic energy to the second one. 'Moderators', such as graphite and water, can be used to slow neutrons down.

Each nuclear fission event liberates 200 MeV, that is, millions of times more energy than that released in chemical processes such as burning fossil fuels. This process occurs with the emission of some neutrons, which in turn induce further fissions, thereby initiating a chain reaction, an idea first proposed by the Hungarian physicist Leó Szilárd.

When Fermi calculated whether a chain reaction could be sustained in a homogeneous mixture of uranium and graphite, he got a negative answer. That was because most neutrons produced by the fission of uranium-235 were absorbed by uranium-238 before inducing further fissions. The right approach, as suggested by Szilárd, was to use separated blocks of uranium and graphite. Fast neutrons produced by the splitting of uranium-235 in the

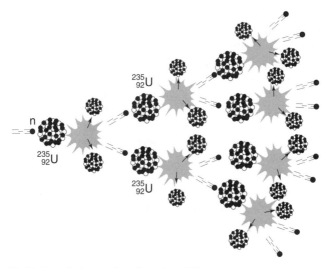

10. Nuclear chain reaction of uranium-235

uranium block would slow down, in the graphite block, and then produce fission again in the next uranium block.

A minimum mass – the *critical mass* – is required to sustain the chain reaction; furthermore, the material must have a certain geometry. The *fissile* nuclides, capable of sustaining a chain reaction of nuclear fission with low-energy neutrons, are uranium-235 (natural isotopic abundance 0.720%), uranium-233, and plutonium-239. The last two don't occur in nature but can be produced artificially by irradiating with neutrons thorium-232 and uranium-238, respectively – via a reaction called *neutron capture*. Uranium-238 (99.27%) is *fissionable*, but not *fissile*. In a nuclear weapon, the chain reaction occurs very rapidly, releasing the energy in a burst.

Fermi and his team evaluated the best uranium graphite arrangements and constructed the first nuclear reactor in a squash court located under the University of Chicago football stadium, and hence named the 'Chicago Pile Number 1'. It consisted of a 2 x 2 x 4 m³ pile of alternating blocks of graphite and uranium. The 57 layers of material included 6 tons of uranium metal, 40 tons of uranium oxide, and 380 tons of graphite. Rods of cadmium (an element that strongly adsorbs slow neutrons) were installed over the pile to control the chain reaction. A 'suicide squad' was also on top of the reactor, ready to throw buckets of a cadmium-salt solution into the pile to interrupt the chain reaction in case of emergency.

At 3:25 pm on 2 December 1942, Fermi gave the order to pull the cadmium rods out slowly. The neutron counter started ticking faster and faster, announcing the first man-made chain reaction: a good time to open the bottle of Chianti that Eugene Wigner (a Nobel Prize winner, in 1963) had brought to the site.

It was not the first time that this phenomenon had occurred on our planet.

About 2 billion years ago, a natural self-sustaining nuclear chain reaction occurred in the Oklo uranium deposits in Gabon, West Africa, moderated by natural water. At least 17 nuclear reactors became critical in this uranium deposit, each operating at 20 kW. The discovery of this phenomenon was made in the 1970s when French scientists found uranium-235 to uranium-238 ratios of 0.717%, slightly lower than the normal natural values of 0.720%, in the Oklo deposit.

According to nucleosynthesis theories that we will discuss in Chapter 7, the original abundances of uranium-235 and uranium-238 were 34% and 66%, respectively. Since then, they decayed according to their half-lives: U-235 ($T_{1/2}$ = 0.704 billion years) is depleted much faster than U-238 ($T_{1/2}$ = 4.468 billion years). During the Precambrian, uranium-235 had not yet decayed to the present level of 0.720% but was enriched to 3–4%. Hence,

Oklo's uranium deposit had a level of uranium enrichment similar to that used in modern light-water reactors.

The study showed that the Oklo reactor had run for a few hundred thousand years. During this period, more than a tonne of plutonium and other transuranic elements were produced in the deposit, together with five tonnes of fission products. Since then, the radioactive products have decayed into stable elements. Isotopic abundances of neodymium, rubidium, and other elements showed anomalies that could be explained with the ancient fission process.

Interestingly, the study of Oklo produced useful insights into the fate of radioactive products in geologic systems, and delivered important information for designing storage facilities suitable for the long-term disposal of all the nuclear waste produced by the many reactors constructed after the Chicago Pile Number 1.

Today, the use of Enrico Fermi's atomic pile for the peaceful production of energy is still controversial. Issues related to safety, non-proliferation, the environment, and nuclear waste still concern the public and decision-makers, hindering the widespread use of nuclear power for electricity generation.

Nuclear fission reactors

The basic components of nuclear power reactors, fuel, moderator, and control rods, are the same as in the first system built by Fermi, but the design of today's reactors includes additional components such as a pressure vessel, containing the reactor core and the moderator, a containment vessel, and redundant and diverse safety systems. Recent technological advances in material developments, electronics, and information technology have further improved their reliability and performance. The reactor core typically includes assemblies of fuel rods, each containing pellets of uranium oxide (UO_2), in some cases mixed with plutonium oxide (PuO_2). The moderator to slow down fast neutrons is sometimes still the graphite used by Fermi, but water, including 'heavy water' – in which the water molecule has a deuterium atom instead of a hydrogen atom – is more widely used. Control rods contain a neutron-absorbing material, such as boron or a combination of indium, silver, and cadmium.

To remove the heat generated in the reactor core, a coolant – either a liquid or a gas – is circulating through the reactor core, transferring the heat to a heat exchanger or directly to a turbine. Water can be used as both coolant and moderator. In the case of boiling water reactors (BWRs), the steam is produced in the pressure vessel. In the case of pressurized water reactors (PWRs), the steam generator, which is the secondary side of the heat exchanger, uses the heat produced by the nuclear reactor to make steam for the turbines. The containment vessel is a one-metre-thick concrete and steel structure that shields the reactor.

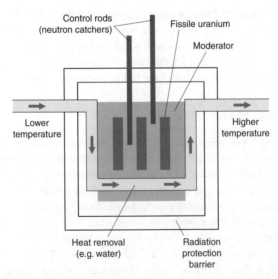

11. Nuclear reactor working principle

Nuclear energy contributed 2,518 TWh of the world's electricity in 2011, about 14% of the global supply. As of February 2012, there are 435 nuclear power plants operating in 31 countries worldwide, corresponding to a total installed capacity of 368,267 MW (electrical). There are 63 power plants under construction in 13 countries, with a capacity of 61,032 MW (electrical). Nuclear technology progressed rapidly after the first experimental reactors.

On 20 December 1951, the world's first electricity-generating nuclear power plant, the Experimental Breeder Reactor I (EBR-I), in Idaho, USA, produced enough electricity to illumine four light bulbs. The nuclear power plant APS-1, the first commercial plant, was connected to the electricity grid at Obninsk, Russia, on 26 June 1954, with an electrical output of 5 MW.

Energy experts predict that nuclear energy will have a heightened role in the near future as part of the global energy mix, as countries pursue greenhouse gas emission reduction targets. It is critical to correct the world's energy imbalance, with 3 billion people still lacking access to electricity. Consider that almost 3 out of 4 of the people living in Africa don't have electricity, a key ingredient for development.

Many of the new power reactors are being constructed or planned in developing and emerging countries, notably in China and India. However, the recent Fukushima Daiichi accident in Japan has forced several countries to slow down or postpone their national nuclear energy programmes. In 2011, Germany has decided to completely stop its nuclear programme by 2022. In Italy, in a national referendum, more than 90% of people voted against the government proposal to re-start a nuclear power programme.

Nuclear power reactors have been classified into four generations of technological development.

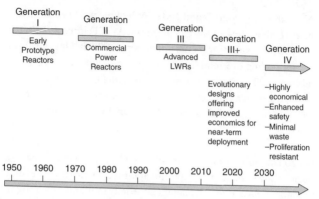

12. Evolution of nuclear power reactors

Generation I refers to early prototype reactors constructed in the 1950s and early 1960s. They generally used natural uranium fuel and graphite as moderator.

Generation II includes commercial reactors built between the late 1960s and the mid-1990s. Most of the reactors in operation today belong to this category. They generally use enriched uranium fuel (uranium-235 isotopic concentration increased from its natural level of 0.7% to 3–4%) and are moderated and cooled by water.

This generation includes the following reactors (at the time of writing, 9 March 2012):

- The pressurized water reactor (PWR) is used in more than 272 nuclear power plants in 26 countries, including the USA, France, Japan, Russia, and China, and in hundreds of ships for propulsion. It was originally designed as a power plant for submarines. Water is used as both moderator and coolant. It has a primary cooling circuit flowing through the reactor under high pressure, and a secondary circuit, which generates steam to drive the turbine and produce electricity.

- The boiling water reactor (BWR) is used in almost 84 power plants in the USA, Japan, Sweden, Finland, Germany, India, Mexico, Spain, Switzerland, and Taiwan/China. It is like PWR, but has a single circuit with cooling water.

- The pressurized heavy water reactor (PHWR) is used in 47 power plants in Canada, Romania, Korea, China, India, and Pakistan. The fuel is cooled by a flow of heavy water under high pressure in the primary cooling circuit. As in the PWR, the primary coolant produces steam in the secondary circuit to drive the turbines.

- The gas-cooled reactor (GCR) is used in 16 UK power plants. It uses carbon dioxide as a coolant and graphite as moderator.

- The light water graphite-moderated reactor (RBMK) is used in 15 Russian power plants (plus one under construction). It has been developed from plutonium production reactors. It has pressure tubes running through the graphite moderator and is cooled by water.

Generation III comprises advanced reactors built since the mid-1990s. They are design upgrades on the Generation II reactors, incorporating improvements in fuel utilization, thermal efficiency, and safety. Their design is standardized to cut capital costs and maintenance. More than 10 advanced designs have been developed. They include reactors that evolved from PWR and BWR, such as the advanced boiling water reactor (ABWR), the European Pressurized Water Reactor (EPR), and the System 80+. There are also more innovative designs such as the pebble bed modular reactor, which is cooled by helium.

Generation IV reactors are still only conceptual, and most of them will not be operational until 2030. They involve innovative approaches to address various concerns about economic competitiveness, safety, security, waste, resistance to diversion of materials for weapons proliferation, and security from terrorist attacks.

The Generation IV International Forum (GIF) is an international initiative aiming to select and develop the new reactor systems.

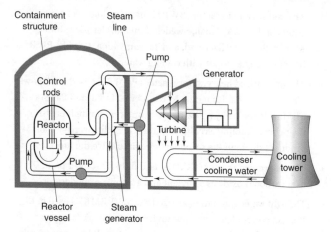

13. Pressurized water reactor (PWR). Water is used as both moderator and coolant

It involves Argentina, Brazil, Canada, China, France, Japan, Russia, the Republic of Korea, South Africa, Switzerland, the UK, the USA, and the EU (Euratom). A similar discussion is also promoted by the IAEA within the International Project on Innovative Nuclear Reactors (INPRO), an initiative that involves 35 IAEA member states and the European Commission.

GIF and INPRO consider a number of innovative concepts, based on different fuel cycles and on the use of thermal, fast, and epithermal neutrons.

Fast reactor and closed fuel cycle concepts

The fuel cycle includes mining and processing of uranium, enrichment (if needed) of uranium-235, production of the nuclear fuel, use of the fuel in the reactor, removal and storage of the used fuel, and possibly reprocessing, to separate the recyclable uranium and plutonium. When the spent fuel is reprocessed, the cycle is defined as *closed* (in Generation IV concepts, also the actinides are recycled and only the fission products are stored and disposed of). The cycle is defined as *open* when the spent fuel is stored.

Fast reactors rely on fast neutrons to cause fission; they don't require a moderator. The fuel of fast reactors consists of plutonium-239 and uranium-238. Fast reactors allow the conversion of non-fissile uranium-238 into fissile plutonium-239 – the material used for the fission process, that is, during reactor operation more fissionable fuel is produced than consumed. The coolant is a liquid metal, such as sodium, used to avoid neutron moderation and to allow very efficient heat transfer.

The conventional fast reactors built so far have a blanket of uranium (called a fertile blanket) surrounding the core in which the plutonium-239 is produced. The blanket is reprocessed to recover the plutonium, in order to be used again as fuel. Hence, both plutonium consumption and production occurs, in a *closed* cycle system.

Fast reactors minimize neutron-capture reactions and maximize fission in uranium and plutonium. They can also be used to fission americium, protactinium, and other actinides produced from ordinary thermal reactors, hence reducing the amount of long-lived radionuclides in high-level waste.

GIF has selected five reactors systems that employ a closed fuel cycle: gas-cooled fast reactors, lead-cooled fast reactors, molten salt reactors, sodium-cooled fast reactors, and supercritical water-cooled reactors.

Accelerator-driven systems (ADS)

ADS systems combine a subcritical nuclear reactor core with an ion accelerator providing extra neutrons. Proposed designs include linear particle accelerators (LINACs) or circular particle accelerators (cyclotrons). Both can produce milliamperes of proton beam currents at energies of 1 GeV. The high-energy

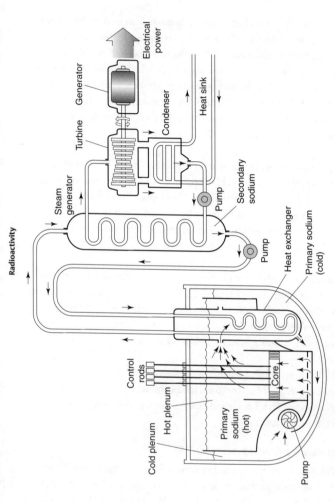

14. Scheme of a sodium-cooled fast reactor

protons hit a high-atomic-number target (tungsten, lead, etc.) and produce neutrons through the spallation process (a high-energy nuclear reaction involving the disintegration of the target). These additional neutrons make the reactor critical. Hence, the chain reaction would stop when the proton beam is turned off, giving ADS systems considerable safety advantages. However, design and operation of the plant would be very complex.

Thorium-based reactor and fuel cycle systems

As we have seen before, thorium is three to four times more abundant than uranium in the Earth's crust. However, it does not contain fissile isotopes. Natural thorium contains only the 'fertile' isotope thorium-232, so it has been combined with fissile uranium-235 and plutonium-239 in power reactors, for conversion to fissile uranium-233. Since the beginning of the nuclear era, there has been much interest in the use of thorium in fuel cycles to expand the availability of nuclear fuel, with ongoing programmes, mostly in India. Unfortunately, complicated reprocessing and refabrication processes are required to overcome the problem of high gamma radiation coming from the short-lived uranium-232 (always associated with uranium-233).

Yet, thorium-based reactors are now considered an important option for innovative Generation IV systems. Apart from its natural abundance, which would increase the long-term sustainability of nuclear power, thorium has other advantages. The thorium fuel cycle generates much lower levels of long-lived actinides, minimizing the radiotoxicity of the spent fuel. Finally, thorium-based reactors would offer intrinsic resistance to nuclear weapons proliferation due to the presence of uranium-232 and its strong gamma-emitting products.

Small modular/transportable reactors

As early as the 1980s the US Air Force was planning to construct small transportable reactors. The concept of small modular designs has been resurrected as part of the next generation of nuclear reactors.

The future small reactors will be based on integral systems including reactor core, control systems, safety systems, and a steam generator, housed in sealed containers that can be transported to sites by ship or heavy transport. At the end of their life, they would be returned to the manufacturer for recycling the fuel. This would prevent the user from opening the reactor to use the generated plutonium for nuclear weapon fabrication.

They will include both thermal and fast reactors, designed to deliver limited power, from 10 to 300–400 MW. NUSCALE, the 45-MW light water reactor developed by Oregon State University, can be part of multiple units whose number can be increased when energy demand expands.

The models based on fast reactors, such as the lead-cooled reactor SSTAR (Small, Sealed, Transportable, Autonomous Reactor) model developed at Lawrence Livermore Laboratory, would operate for 30 years or more without refuelling. SSTAR, which could produce from 10 to 50 MW of electricity, weighs 500 tonnes with a diameter of 3 metres and a height of 15 metres.

Also, Russia is developing similar concepts, using its long experience in the use of small reactors in icebreakers. These transportable reactors, like the KLT-40S model, could be used on barges for supplying electricity to remote regions. One unit can produce a power of 35 MW to be used either for electricity or for producing heat for desalination. They can also be used for some years without refuelling.

In spite of the many concepts being discussed and of their benefits, fission reactors of the fourth generation have still a long way to go. Alternative nuclear energy systems, based on fusion reactions of light nuclei, will not be available in the foreseeable future either.

Nuclear fusion

Nuclear fusion is believed to be the long-term solution to the energy problem of humankind. It is the same process that has kept the Sun shining for the past 5 billion years, and the Sun is only a middle-aged star.

'What is possible in the Cavendish Laboratory may not be too difficult in the Sun.' This statement is attributed to the English astrophysicist Sir Arthur Eddington, who first suggested, in 1920, that the energy from the Sun was the result of the transmutation of hydrogen into helium. He based his theory on the discovery, made the same year by the American spectrometrist Francis William Aston, that four hydrogen nuclei were heavier than a helium nucleus. According to the famous mass-energy relation that Einstein had revealed in 1905, the nuclear fusion of hydrogen could power the Sun for 100 billion years. That was much longer than the 20 to 100 million year span calculated 100 years earlier by physicists such as Kelvin and von Helmholtz, who had assumed that the only source of solar energy was the gravitational energy accumulated during its accretion.

In his 1939 paper 'Energy Production in Stars', the German-American physicist Hans Bethe discussed all the basic nuclear processes driving the hydrogen fusion into helium inside the Sun and other stars. Bethe calculated the temperature at the centre of the Sun and obtained a value that was within 20% of what we presently consider the correct value (16 million K). He was also able to find a relationship between the mass and the luminosity of stars, which agreed with astronomical observations. Bethe was awarded the Nobel Prize in 1967 'for his contributions to the theory of nuclear reactions, especially his discoveries concerning the energy production in stars'.

Among all the possible reactions, Bethe selected the two most important for the production of energy in the Sun. The first reaction, called the *p-p* chain, fuses four hydrogen nuclei, producing helium; this is the main source of energy in the Sun and in other relatively small stars. The chain could include different reaction sequences, but the most likely is the following: in the first phase, two hydrogen nuclei fuse to produce a deuterium nucleus; in the second phase, deuterons fuse with protons to produce helium-3; the third phase can proceed via different reactions, all of which lead to the production of helium-4 nuclei. The net result of the *p-p* chain is the fusion of four hydrogen nuclei into a single helium-4, with the release of an amount of energy equal to the difference between the masses of the four hydrogen-1 nuclei and the helium-4 nucleus. The second mechanism (called the CNO cycle – which stands for carbon, nitrogen, and oxygen) is based on the use of these elements as catalysts in a chain reaction that produces the same effect: four hydrogen nuclei produce a helium nucleus, together with some other particles.

Gravitation has concentrated all the hydrogen available in the early universe over billions of years, creating large masses that have cores at extremely high temperatures and densities – the right conditions for nuclear fusion. The hydrogen nuclei can overcome electrostatic repulsion and fuse under the effect of the nuclear force. In the Sun, 600 million tonnes of hydrogen nuclei are turned into helium nuclei every second.

So how can we possibly reproduce these conditions to obtain the energy we need, at our much smaller scale, in a controlled process?

Scientists have decided that the best nuclear reaction that we can use for energy production is the fusion between hydrogen-2 (deuterium) and hydrogen-3 (tritium), which is the most efficient in terms of energy produced. The reaction yields in fact 17.6 MeV of energy for each fusion reaction. Yet, a temperature of more than 40 million K (higher than that at the centre of the Sun) must be achieved to overcome the Coulomb barrier created by the repulsion of the positive charges of the two nuclei.

At the temperatures required for fusion, the hydrogen mass becomes a plasma, a state of matter in which atomic electrons are separated, creating an electrically charged mixture of positive ions and electrons. Like gas, a plasma does not have a definite shape or volume, but being charged, it can be shaped by magnetic fields. It was first identified by Crookes in 1879, and was dubbed 'plasma' by the American physicist Irving Langmuir in 1928.

Magnetic fields are used to control and contain the high-temperature plasma inside a doughnut-shaped magnetic confinement device, the so-called tokamak. The deuterium and tritium of the plasma fuse to produce helium nuclei and neutrons, releasing energy. Helium nuclei are confined within the magnetic field of the tokamak, while neutrons abandon the plasma, carrying 80% of the energy, which is released as heat to the walls of the confinement system.

Fusion experiments started in the 1930s, but the first tokamak, T1, built at the Kurchatov Institute in Moscow, was successfully operating in the USSR only in 1968.

Since then, more than 200 tokamaks have been built, including the Joint European Torus (JET), the Japanese JT-60, and others in the United States. The International Thermonuclear Experimental Reactor (ITER) is the largest and most advanced device that is being built on the base of this concept. ITER is being developed in Cadarache, France, by a consortium of nations including China, members of the European Union, India, Japan, Korea, Russia, and the United States, representing over half of the world's population. ITER's tokamak will be constructed by 2018, and the first plasma will be produced in 2019. ITER will continue with DEMO, a prototype fusion power plant that will become operational in

2030, with the possibility of putting fusion power into the grid in 2040.

Is the fuel for deuterium–tritium fusion easily available? Deuterium is present in the hydrogen of seawater at the level of one part in 5,000, so this is not a problem. The main challenge is that tritium has a short half-life (12.3 years) and cannot be found in nature. It can be produced by bombarding lithium-6 in a reactor using the nuclear reaction:

$$_3^6 Li_3 + {}_0^1 n_1 \rightarrow {}_2^4 He_2 + {}_1^3 H_2 + 4.8 \ MeV$$

A second concept being considered, for controlled fusion experiments, is based on 'inertial confinement', in which the reaction between deuterium and tritium is ignited by a high-energy pulse produced by intense laser beams or nuclear bombs.

This concept is the basis of nuclear weapons in which the compression and the high temperature needed to trigger fusion comes from the detonation of a fission bomb, based on enriched uranium or plutonium. The first so-called 'hydrogen' or 'thermonuclear' bomb was detonated on 1 November 1952 in the Marshall Islands, by the United States. A megaton-sized fusion bomb was exploded by the USSR in 1953. The largest hydrogen bomb ever exploded released as much energy as 55 million tons of TNT, corresponding to more than 4,000 Hiroshima bombs. Most of the bombs in the US arsenal are hydrogen bombs, as they are more compact and effective. Thermonuclear bombs can also be made using deuterium and lithium-6, combined in a solid target of lithium deuteride. During the first phase of the explosion, neutrons from the fission of plutonium react with lithium, according to the nuclear reaction shown above, producing tritium. The fusion between tritium and deuterium proceeds during the second phase. The neutron bomb is a special fusion device that alarmed the public a few years ago. It was created by minimizing the primary fission stage and maximizing the fusion stage that is based on a neutron-producing reaction. It is particularly sinister because it could be used to kill people, with only minimum contamination of the environment and minor damage to infrastructure and constructions.

The two main projects promoting inertial confinement for energy production are the Laser Megajoule in Bordeaux, France, and the National Ignition Facility at the Lawrence Livermore Laboratory in the United States. In reality, these facilities are also important to validating the computer programs used to test the development of nuclear weapons (overcoming the limitations of the international treaties on nuclear testing).

The deuterium–tritium pellet of fuel is contained in a small cylinder of gold. One megajoule laser pulse of ultraviolet light was recently delivered to such a target at Livermore using 192 powerful lasers. This system can deliver a power of 500 trillion watts, 1,000 times the power generated for producing electricity in the United States, but this power is produced only for four-billionths of a second, releasing an energy of 1.8 megajoules in a volume of one cubic millimetre. This is only the energy that can be produced by about 50 millilitres of petrol. These fusion cycles would have to be repeated at a very high rate to compete with existing power systems.

US scientists expect to achieve a self-sustaining fusion reaction with energy gain within the next two years, but it will probably take 30 years to produce electricity in a useful way. Since the first nuclear fusion, more than 60 years ago, many have argued that we need at least 30 years to develop a working fusion reactor, and this figure has stayed the same throughout those years. For the time being, Laser Megajoule and the National Ignition Facility remain experimental facilities that produce microscopic thermonuclear explosions, useful for testing new nuclear weapons.

In the meanwhile, other ideas are being explored, including the fusion of helium-3. This is an attractive option as this fuel is non radioactive, and there is no production of neutrons and radioactive products. Unfortunately, helium-3 is extremely rare on Earth, but it could be mined on the Moon, where this gas has been stored by the solar wind during geologic times. Helium-3 is also a by-product of tritium used in nuclear weapons.

Chapter 3
Food and water

When the Hungarian chemist George de Hevesy joined Ernest Rutherford's group in 1911, he became suspicious that the landlady at his boarding house was serving the previous day's leftovers at dinnertime. To prove his theory, he introduced a small amount of radionuclides in the food he had not eaten. Analysing the food served the next day in his laboratory, he indeed confirmed his suspicions. In the following years, he obviously applied his scientific creativity more seriously, receiving the Nobel Prize in Chemistry in 1943 'for his work on the use of isotopes as tracers in the study of chemical processes'.

In agriculture, the methods developed by de Hevesy have been used extensively to label nucleic acids with radionuclides to characterize mutations and assist in genetic selection.

It is well known that ionizing radiation is also used to improve many properties of food and other agricultural products. For example, gamma rays and electron beams are used to sterilize seeds, flour, and spices. They can also inhibit sprouting and destroy pathogenic bacteria in meat and fish, increasing the shelf life of food. A dose between 50 and 150 gray inhibits sprouting in onions, garlic, and potatoes. A dose between 1,000 and 4,000 gray can extend the shelf-life of strawberries and other fruit. It takes a dose up to 7,000 gray to kill bacteria such as salmonella in meat and fish. The protection of spices from micro-organisms and insects needs up to 30,000 gray.

More than 60 countries allow the irradiation of more than 50 kinds of foodstuffs, with 500,000 tons of food irradiated every year. About 200 cobalt-60 sources and more than 10 electron accelerators are dedicated to food irradiation worldwide.

Genetic mutations

Darwin's ideas on evolution by natural selection were inspired by selective breeding. He performed many experiments on plants grown in his own garden at Downe, near London.

In *On the Origin of Species*, he wrote:

> Man does not actually produce variability; he only unintentionally exposes organic beings to new conditions of life, and then nature acts on the organisation, and causes variability. But man can and does select the variations given to him by nature, and thus accumulate them in any desired manner.

Darwin published his theory 37 years before the discovery of radioactivity cleared the way to artificially mimicking natural spontaneous mutations.

With the help of radiation, breeders can increase genetic diversity to make the selection process faster. The spontaneous mutation rate (number of mutations per gene, for each generation) is in the range 10^{-8} –10^{-5}. Radiation can increase this mutation rate to 10^{-5}–10^{-2}.

Gamma rays and other types of ionizing radiation can be used to induce mutations in plants to increase 'germplasm' (living tissue from which new plants can be grown) and enhance variability to help breeders improve crop quality. New varieties with new qualities such as higher yield and disease resistance are normally chosen in selective breeding programmes. Specific characteristics, including taste and size, can be changed without affecting the general nutritional properties of the plant. The process mimics natural radiation-induced mutations, and is thus quite distinct from the more controversial engineering of genetically modified organisms, which occurs via the introduction of foreign genetic components.

Radiation was first applied to plant breeding in the 1920s, when the American geneticist Lewis Stadler identified mutations induced by X-rays in maize and barley. He verified that the mutations resembled those occurring spontaneously. Another American geneticist, Hermann Muller, a political activist who campaigned against nuclear testing, conducted similar experiments.

The early experiments produced curious mutants, such as plants possessing leaves with white stripes, but scientists soon found that this method produced more useful mutations when applied to large numbers of seeds.

More than 2,600 mutant varieties of more than 160 plant species are now listed in the FAO/IAEA (FAO is the UN organization for food and agriculture) Mutant Variety Database. Among them, we find a mutant rice strain extremely tolerant to salinity. It is being cultivated in Vietnam, and will also be introduced into Bangladesh, India, and the Philippines. A second example is the development of date palms that are tolerant against Bayoud disease, a major limiting factor for this culture in North Africa. Several useful mutant varieties of wheat, barley, cassava, sunflower, banana, sesame, grapefruit, and linseed are being bred.

Plant breeding is expected to help less developed countries to cope with climate change impacts, including the contraction of arable land and the drying up of drinking water supplies.

Water

More than a million people in developing countries lack access to clean water. Just over 40% of Africans don't have access to potable water. Experts predict that without urgent action, more than 60% of people will face water shortages by 2025.

Knowledge of groundwater sources, age, and recharge rate, together with the sources of possible pollution, is of course critical to managing the hydrological cycle. Radioisotope techniques can contribute to the global quest for water security. Several stable isotopes, along with radiogenic and cosmogenic radionuclides, are used as tracers and clocks. They include the isotopes of hydrogen, oxygen, and carbon, long-lived radionuclides such as carbon-14 ($T_{1/2}$ = 5,730 years), chlorine-36 ($T_{1/2}$ = 301,000 years), iodine-129 ($T_{1/2}$ = 15.7 million years), and krypton-81 ($T_{1/2}$ = 230,000 years).

Accelerator mass spectrometry (AMS) is the analytical technique of choice for the detection of long-lived radionuclides that cannot be practically analysed with conventional mass spectrometry (MS) or decay counting. AMS uses an ion accelerator as a component of the mass and charge spectrometer. Hence, ion detectors can be used to identify nuclear mass and atomic number of each individual ion extracted from the sample and accelerated to high energies. Molecules are the main limitation in MS, as they cannot be distinguished from the nuclide of interest when they have the same mass. Thanks to the use of an accelerator, molecules can be destroyed by sending the high-energy ions through a foil or a gas where their binding electrons would be stripped off. In conclusion, AMS enables the analysis of isotopic ratios as low as 10^{-15}, a factor 10^6 lower than in most MS systems. In addition, since with AMS, the atoms are directly counted, rather than detected by measuring the radiation resulting from their decays, the measurement sensitivity is unaffected by the half-life of the isotope. Hence, compared to the decay-counting technique, the efficiency of AMS in detecting long-lived radionuclides is 10^5–10^9 times higher, the size of the sample required for analysis can be 10^3–10^6 times smaller, and the measurement can be performed 100 to 1,000 times faster.

Long-lived cosmogenic radionuclides provide unique methods to evaluate the 'age' of groundwaters, defined as the mean subsurface residence time after the isolation of the water from the atmosphere. This age is calculated by measuring the decrease of the radionuclide concentration from the known value assumed by

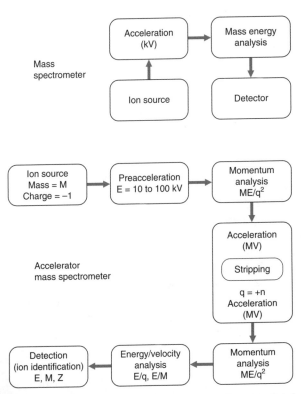

15. Schematics of an accelerator mass spectrometer (a conventional mass spectrometer is also shown for comparison)

the surface water that moved downward (through the complicated hydrologic processes contributing to groundwater recharge).

Scientists can date groundwater more than a million years old, through chlorine-36, produced in the atmosphere by cosmic-ray reactions with argon. But the method is complicated by uncertainty about the concentration of the radionuclide when it enters the 'aquifer' (a geologic formation yielding significant quantities of water to wells and springs).

The Great Artesian Basin, one of the largest groundwater basins in the world, with 64,900 cubic kilometres of groundwater, is a vital source of freshwater for inland Australia. Chlorine-36 has been analysed to date its waters, which are more than a million years old. The samples had chlorine-36 to chlorine ratios down to 10^{-16}. It was necessary to use the large tandem Van de Graaff accelerator at the Australian National University in Canberra to carry out these analyses. The information was used to work out flow conditions and identify the areas of groundwater recharge. The application of this method is complicated by the subsurface input from uranium-produced chlorine-36, which contaminates the cosmogenic chlorine-36.

Another groundwater chronometer is krypton-81, produced in the stratosphere by high-energy cosmic rays. Krypton-81 is not hampered by the problems besetting the use of chlorine-36 and is now measurable with sufficient sensitivity using AMS and laser-based atom-counting methods. The analysis of chlorine-36 and krypton-81 has been used to study climatic effects on the Nubian Aquifer in the Sahara Desert during the last million years, showing that the method can be applied to a broad range of hydrological problems.

Arsenic can reach dangerously high concentration in some groundwater systems. Chronic arsenic poisoning is a major problem in some regions of Bangladesh and India. It has been estimated that

80% of Bangladesh is affected by arsenic, threatening the health of 40 million people. Scientists use radioactivity to study the sources and processes of arsenic mobilization and concentration. Tritium and carbon-14 are used to evaluate groundwater ages and their connection to arsenic mobilization. The presence of tritium and radiocarbon deriving from bomb testing shows that some of the arsenic-contaminated waters are relatively young. On the other hand, older aquifers, dating back to the Pleistocene, are characterized by low levels of arsenic and would be a much better source of water for people of these regions.

Zapping the pest

Pests wreak havoc on humans, livestock, and crops, but the insecticides used widely can harm the environment. Pesticides can kill bees and other beneficial insects, contaminate farm workers, water, and soils. A good alternative is the sterile insect technique (SIT).

The SIT is based on the sterilization of mass-bred pest males using a dose of gamma radiation from a cobalt-60 source. The insects must be given a precise radiation dose, enough to sterilize them without compromising their health. After the irradiation, the sterile pupae are dropped from an aeroplane in the infested area. The sterile males mate many times with wild females, which mate only once. If enough males are released, the overall population of insects plummets, and the pest can sometimes be eradicated from an area. The method works well in isolated areas such as small islands.

Edward Knipling developed the SIT method in 1938 while working for the US Department of Agriculture in Texas. His aim was to hit the screw worm fly, a pest lethal to livestock. The first tests of SIT were carried out in the early 1950s in Sanibel Island, Florida. During the following half century, the screw worm was eradicated from North and Central America.

SIT is now used widely to eradicate or control pests including tsetse fly, horn fly, Mediterranean fruit fly, and onion fly. The tsetse fly is a big problem for some African countries where it acts as a vector of the so-called 'sleeping sickness'. This disease affects livestock and the human population and constitutes a major hindrance to economic development. United Nations agencies are supporting global programmes to eradicate tsetse fly and other pests.

In the past few years, the IAEA has been promoting intensive programmes to apply SIT to control mosquitoes carrying malaria. In particular, researchers are targeting *Anopheles arabiensis*, a powerful vector of malaria in sub-Saharan Africa. Malaria kills a child in the world every 45 seconds. Almost 90% of these child fatalities are in Africa.

Chapter 4
Radiation and radioactivity in medicine

More than 20 ambulances with portable X-ray equipment, nicknamed 'petite Curies', and 200 stationary units were deployed on the front line in the First World War. Marie Curie herself supported this novel medical application of atomic science with her Nobel Prize money. X-rays were critical to diagnosing bone fractures and finding embedded bullets in wounded soldiers.

X-ray medical departments had been set up since the last years of the 19th century. One was at the Glasgow Royal Infirmary where, under the guidance of Dr John Macintyre, the first X-ray image of a kidney stone was taken in 1896.

During the first decades of the 20th century, medical examinations were performed by irradiating the body part of interest with X-rays and collecting the transmitted radiation on a film. The patients themselves held the film cassettes; the required exposure time could add up to more than 10 minutes. Present systems need only milliseconds, delivering to the patient a dose hundreds of times smaller. X-ray overexposure was a common problem for early X-ray practitioners. They often had to have their fingers amputated after developing malignant ulcers.

By the early 1900s, doctors were using contrast media, like compounds of barium and iodine, which were injected into the body to visualize blood vessels, the gastro-intestinal system, and other internal organs. Further advances in X-ray radiography were the development of fluorescent screens which could be used to obtain real-time moving images.

X-ray intensifiers were invented in the 1950s and could be used with a television camera to display X-ray movies. It was the beginning of angiography, the technique for imaging blood vessels and the heart.

The advent of digital technology and computers in the 1970s revolutionized medical imaging. These advances affected all existing medical-imaging methods previously based on analogue systems, including angiography. The main advantage of digital technology is that images can be enhanced with computer techniques, archived efficiently, and sent via the Internet, allowing for remote medical diagnosis.

A special invention of this period is computerized axial tomography (CAT), developed in 1972 by Sir Godfrey Hounsfield, who received the Nobel Prize in Physiology and Medicine in 1979, together with the South African physicist Allan Cormack. CAT devices are based on the use of multiple radiographies taken from different angles around a single axis of rotation. In a conventional CAT, the X-ray generator and detector are rotated around the part of the body to be imaged. X-ray scans are stored in the memory of a computer and 3-D images, including slices of internal parts of the body, are then reconstructed using mathematical algorithms called 'filtered back projections'. They are based on methods developed in 1917 by the mathematician Johann Radon.

Synchrotron radiation has also been used for mammography in clinical application at the ELETTRA accelerator in Trieste, Italy. Synchrotron X-rays have some special properties that allow high-resolution 'phase contrast' imaging – based on refraction effects of X-rays – resulting in improved image quality and reduced radiation doses.

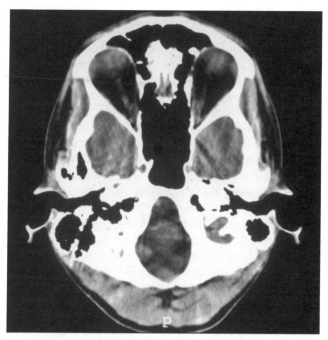

16. X-ray computed tomography of a human brain

Radionuclide imaging was developed in the 1950s using special systems to detect the emitted gamma rays. The gamma-ray detectors, called gamma cameras, use flat crystal planes, coupled to photomultiplier tubes, which send the digitized signals to a computer for image reconstruction. Images show the distribution of the radioactive tracer in the organs and tissues of interest. This method is based on the introduction of low-level radioactive chemicals into the body. It has an important use, for example, for evaluating cardiac disorders. The dose to the patient is comparable to that of similar X-ray analyses, but the radiopharmaceuticals remain in the body for some time, and can trigger alarms in airports or public events where security forces monitor the illicit trafficking of radioactive materials.

More than 100 diagnostic tests based on radiopharmaceuticals are used to examine bones and organs such as lungs, intestines, thyroids, kidneys, the liver, and gallbladder. They exploit the fact that our organs preferentially absorb different chemical compounds. For example, the diagnosis of hyperthyroidism exploits the high accumulation of iodine in the thyroid. Other applications include diagnosis of cardiac stress, metastatic growths in bones, and blood clots in lungs. Many radio-pharmaceuticals are based on technetium-99m (an excited state of technetium-99 – the 'm' stands for 'metastable' – that decays emitting a gamma ray, with a $T_{1/2}$ = 6 hours). This radionuclide is used for the imaging and functional examination of the heart, brain, thyroid, liver, and other organs. Technetium-99m is extracted from molybdenum-99, which has a much longer half-life and is therefore more transportable. It is used in 80% of the procedures, amounting to about 40,000 per day, carried out in nuclear medicine. Other radiopharmaceuticals include short-lived gamma-emitters such as cobalt-57, cobalt-58, gallium-67, indium-111, iodine-123, and thallium-201.

In SPECT (single-photon emission computed tomography), the gamma-ray camera rotates, collecting images at different angles that are used for 3-D reconstructions. One tracer used in SPECT for cardiac stress tests is thallium-201 ($T_{1/2}$ = 73 hours), which emits gamma rays of 135 and 167 keV. This radionuclide has now largely been replaced by technetium-99m. Other radionuclides include iodine-123 (13 hours), which shows ischemic spots in the myocardium, and gallium-67 (78 hours), used to identify points of acute inflammatory cardiomyopathy.

Positron emission tomography (PET) is an advanced imaging technique used for medical diagnostics. The concept was developed in the 1950s, but was only applied in the early 1970s. It is based on the positron emission that results from the decay of radionuclide tracers such as fluorine-18, gallium-68, iodine-124, carbon-11, nitrogen-13 ammonia, and oxygen-15. Fluorine-18 is

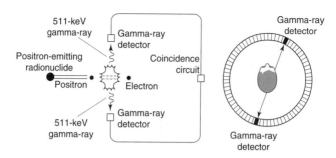

17. Principles of positron emission tomography. The tracer emits a positron that annihilates an electron, emitting two gamma rays, which are in turn picked up by the PET detectors

the most widely used radionuclide, but has a half-life of less than two hours, which requires that the cyclotron used to produce it be near the hospital.

The tracer is introduced into the body as a label of biologically active molecules. Immediately after its production, the positron interacts with an electron and is thus annihilated. The energy of the two masses at rest (511 keV each) is transformed into two gamma rays with the same total energy, travelling in opposite directions to conserve momentum. The total energy produced is related to the annihilated mass by the well-known Einstein formula, $E = mc^2$ (energy equals the mass m times the square of the speed of light c).

The two gamma rays produced by each radionuclide in the organ have enough energy to penetrate outwards the patient's body and be detected simultaneously by the PET detectors. The electronic signals produced by the detectors are digitized and sent to a computer, which produces a 3-D image showing the changes of the radionuclide distribution with time, providing information on the organ's dynamic functioning.

PET can measure how different parts of our brain work while we are alive and fully conscious. The chemical processes governing brain functions can be studied in detail, giving unique information on possible malfunctions. Before PET, doctors could diagnose brain abnormalities and injuries only in post-mortem examinations. While the X-ray CAT shows static structural details of the brain, PET provides a dynamic picture of the brain processes. The information of the PET image is often complemented with an X-ray computed tomography performed on the patient during the same session, allowing for a better localization of the radiopharmaceutical uptake in small volumes of the tissue.

Cancer therapy

In 1898, Henri Becquerel noticed an erythema (inflammation of the skin) on his abdomen. He assumed that it had been caused by a radium tube that he had received as a gift from Marie Curie, which he was keeping in his waistcoat pocket. Pierre Curie suggested he checked his hypothesis by placing the radioactive material in the opposite pocket, and this produced a second erythema. Pierre repeated the experiment on himself. With great foresight, he realized that these substances could be used in medicine, for curing cancer or other diseases.

Henry Danlos, a doctor at the Hôpital St-Louis, Paris, was one of the first who used a radioactive source for medical treatment. In 1901, he used some radium that he had borrowed from Pierre Curie to treat a patient affected by lupus erythematosus, a chronic autoimmune disorder. However, there are even earlier reports of therapeutic uses of radiation. In 1896, the Viennese radiologist Leopold Freund treated a patient with X-rays, which had been discovered just a year before. The patient was a five-year-old girl whose back was covered with large hairy moles. The treatment lasted five years and was successful, but she suffered from the effects of some bad ulcers. She was examined periodically (until she was 75) and found to be relatively well.

Today, cancer treatment with radiation is generally based on the use of external radiation beams that can target the tumour in the body. Cancer cells are particularly sensitive to damage by ionizing radiation and their growth can be controlled or, in some cases, stopped.

High-energy X-rays produced by a linear accelerator (LINAC – see Chapter 2) are used in most cancer therapy centres, replacing the gamma rays produced from cobalt-60. The LINAC produces photons of variable energy bombarding a target with a beam of electrons accelerated by microwaves. The beam of photons can be modified to conform to the shape of the tumour, which is irradiated from different angles.

The main problem with X-rays and gamma rays is that the dose they deposit in the human tissue decreases exponentially with depth. A considerable fraction of the dose is delivered to the surrounding tissues before the radiation hits the tumour, increasing the risk of secondary tumours. Hence, deep-seated tumours must

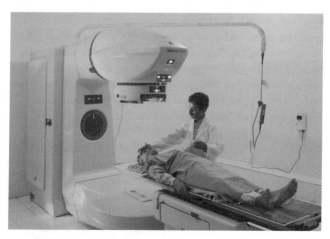

18. System for cobalt-60 therapy

be bombarded from many directions to receive the right dose, while minimizing the unwanted dose to the healthy tissues.

The problem of delivering the needed dose to a deep tumour with high precision can be solved using collimated beams of high-energy ions, such as protons and carbon. The concept of proton therapy was developed by Robert Wilson, one of the physicists of the Manhattan Project, who published in 1946 a paper on the 'radiological use of fast protons', getting a lukewarm reaction from the medical community. Forty years later, the University Medical Centre of Loma Linda in California considered the construction of the first hospital-based proton synchrotron centre. More than 30 centres are now operating, or are under construction, in Europe, the USA, and Japan, using both protons and carbon ions.

Contrary to X-rays and gamma rays, all ions of a given energy have a certain range, delivering most of the dose after they have slowed down, just before stopping. The ion energy can be tuned to deliver most of the dose to the tumour, minimizing the impact on healthy tissues. The ion beam, which does not broaden during the penetration, can follow the shape of the tumour with millimetre precision. Ions with higher atomic number, such as carbon, have a stronger biological effect on the tumour cells, so the dose can be reduced. Ion therapy facilities are still very expensive – in the range of hundreds of millions of pounds – and difficult to operate. There is an ongoing effort to develop cheaper facilities, such as those using pulsed laser for ion acceleration.

The direct delivery of radioactive nuclides to the tumour can in some cases be an alternative to the use of external beams. Some are gamma-ray emitters, like iodine-131, which is used to treat thyroid cancer, and iridium-192 is applied to several kinds of tumours. The radioactive source is inserted through a catheter near the target area (brachytherapy). Auger electron-emitting indium-111 (2.8 days) or beta-emitting rhenium-188 (16.9 hours)

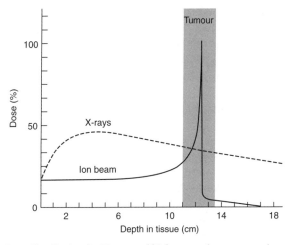

19. Dose distribution for X-rays and high-energy ions penetrating biological tissues

and strontium-89 (50.6 days) are currently used to palliate bone cancer pain.

External radiation doesn't work if the tumour is in metastasis, so treatment has to shift to the cellular level. Approved cancer therapies commonly use beta-emitting radionuclides, such as yttrium-90 (64 hours), but they are unsuitable for working at the level of single cells, as thousands of particles are needed to kill a single cell.

Alpha-emitting radionuclides offer the best option in these cases, since the alpha particles they produce, with energies between 5 and 8 MeV, have a typical range of 50 to 90 microns through living matter, which is comparable to the cell size. Alpha particles have a strong local release of energy, and one single particle is enough to destroy the nucleus of a cell. Alpha particles thus deposit a high amount of energy along their path, thousands of times higher than beta particles. This is normally quantified via the 'linear energy transfer' (LET) parameter, measured in keV per micron. For example, alpha particles of 5 MeV and beta particles of 1 MeV energy have LETs of 95 and 0.25 keV/µm, respectively. Alpha particles deposit most of their energy at the end of their path, with a very high biological impact.

The first clinical trial using alpha radionuclides was based on the use of bismuth-213. Because of its short half-life ($T_{1/2}$ = 45.6 minutes), this radionuclide must be produced in the place where it is delivered to the patient. Bismuth-213 is a daughter product of actinium-225 ($T_{1/2}$ = 10 days) that can be produced from the decay of thorium-229. Recently reported physical trials using bismuth-213 were related to the cure of melanoma, glioblastoma (brain tumour), and myelogenous leukaemia. The energy of alpha particles emitted by bismuth-213 is 8.375 MeV, with a range of 85 microns in human tissue. Their initial LET is 61 keV per micron, while near the end they release nearly 250 keV per micron, hence the required dose will be effectively delivered to the tumour cells.

A group at the University of Washington is testing the alpha-emitter astatine-211, produced using a cyclotron. The half-life of astatine-211 is only 7.2 hours, hence the radioactivity decays rapidly and patients experience fewer side effects. Astatine is loaded into carbon nanotubes of the size of cellular DNA, which deliver the radioactive element to the cells. Antibodies locate the cancerous cells and bind to them, transporting the radioactive atoms to the external membrane (radioimmunotherapy). Clinical trials are under way for ovarian cancer and glioblastoma multiforme.

Boron neutron capture therapy (BNCT) is a special targeted chemo-radiation therapy. With this technique, boron is delivered to tumour cells, emitting charged particles after irradiation with a neutron beam. The stable nuclide boron-10-tagged compound is injected intravenously into the patient, approximately 30 micrograms of boron-10 per gram of tumour. The compound is selected to bind preferentially to tumour cells. Epithermal neutrons (1 eV–10 keV), produced by reactors or accelerators, can be used to irradiate the tumour after having been slowed down in a moderator. After reaching thermal energies (0.025 eV), neutrons react with boron-10 producing alpha particles and lithium-7, which deposit their energy within less than 10 microns from the reaction point. This method has been trialled for glioblastoma cancers in the brain.

An Italian group carried out the clinical application of BNCT on two patients with liver metastasis. The procedure involved injecting the boron-10 compound into the patient, surgically removing the liver, and externally irradiating it with thermal neutrons from a reactor, and finally re-implanting the liver. One of the two patients died soon after the operation, while the second survived 44 months with a reasonable quality of life.

Tracing toxins

Radionuclides have been applied for decades as tracers in biomedicine by incorporating them into specific molecules. In general, radiolabelling is carried out using short-lived radioisotopes so that decay-counting times are reasonable. The development of accelerator mass spectrometry, which we discussed in Chapter 3, made the application of longer-lived radionuclides possible. After revolutionizing archaeology with high-precision radiocarbon dating of bodies that had been dead for thousands of years, AMS is now having a unique impact in biomedicine, measuring radiocarbon in living humans and in rats.

Since AMS does not require waiting until the radioisotope has decayed, the analysis can be done more rapidly and with smaller amounts of radioactivity, reducing the tissue's exposure to radiation. Using 'modern' carbon-14 isotopic concentrations (carbon-14 to carbon-12 ratio of about one part per trillion), approximately 10,000 radiocarbon ions can be counted within a minute. AMS is even more efficient using other long-lived radionuclides of biomedical interest such as aluminium-26 and calcium-41.

The relationship between DNA adducts – carcinogen molecules bound to the DNA double helix – and carcinogen dose has been studied with AMS at the Lawrence Livermore National Laboratory. In one experiment, mice were fed with very low levels of a radiocarbon-labelled carcinogen (PhIP) produced at levels on the parts-per-billion scale in cooked meat. Previous studies with conventional techniques involved doses to mice that were equivalent to the PhIP content of 100 million hamburgers, whereas AMS is able to detect the PhIP content equivalent of just one hamburger. Perhaps not surprisingly, there are important differences in rates of removal for low- and high-dose conditions. More importantly, AMS is able to detect the tiny amounts of PhIP reaching individual organs and to quantify its removal as a function of time after dose.

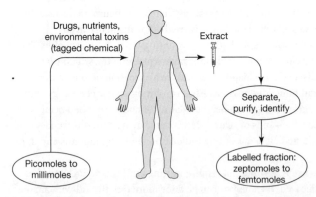

20. The radiotracer method consists of the introduction into the body of a substance with radionuclide-labelled molecules, followed by the measurement of the tracer accumulation in the specific organs or parts of the body. Long-lived radionuclides detected with AMS allow long-term biomedical studies

With these methods, trace amounts of drugs, nutrients, and environmental toxins, from picomoles to millimoles (10^{-12}–10^{-3} moles), can be introduced into the body; the labelled fractions of interest, after separation and purification, correspond to amounts from femtomoles to zeptomoles (10^{-15}–10^{-21} moles). (One mole is an amount of a substance containing as many atoms or molecules as there are atoms in 12 grams of pure carbon-12.)

Several experiments have been aimed at the study of gastro-intestinal and intravenous absorption of aluminium and its metabolism in humans. For example, 100 nanograms (70 Bq) of aluminium-26 were ingested with 100 millilitres of orange juice by six volunteers: 80% of the aluminium was excreted within 10 days, and the rest was excreted slowly. One-millionth of the initial dose remained after 1,000 days.

A group based in Sydney, Australia, used aluminium-26 to measure the uptake of aluminium in rat brains from drinking water, quantifying the time it took aluminium to enter the bloodstream and cross the blood–brain barrier. The work was aimed at studying the association of aluminium with Alzheimer's disease, and to evaluate the risks of using aluminium compounds to treat potable water. The study showed that trace amounts of aluminium-26 from a single exposure directly entered their brain tissue. Hence, uptake of aluminium into the human brain, from aluminium-treated drinking water over a prolonged period of time, may contribute to long-term health consequences for some people.

The metabolism of calcium is the focus of research aimed at understanding bone diseases such as osteoporosis. Calcium is absorbed from food through the intestines with about 30% efficiency. An imbalance between bone accretion and resorption causes osteoporosis. In the past, the short-lived radioisotopes calcium-47 ($T_{1/2}$ = 4.5 days) and calcium-45 ($T_{1/2}$ = 163 days) have been used as tracers. However, the long-term effects of osteoporosis cannot be studied using these calcium radionuclides due to their short half-life. Calcium-41 ($T_{1/2}$ = 104,000 years) is a viable alternative for studies in the long period, but only AMS detection makes its application practical.

In 1990, in a bone resorption study of menopausal women, a volunteer ingested 125 nanograms (320 Bq) of calcium-41. The annual limit of intake for calcium-41 is 100 million Bq. The radiation dose commitment was only 0.42 mSv in 50 years, against a natural radiation background of 2.4 mSv/yr. The AMS analysis requires only 1 millilitre of urine and can be done in less than one hour. In a pilot long-term feasibility study, calcium-41 in urines has been measured for six years from premenopausal to perimenopausal phases. The bone density was measured in parallel. This method is still under evaluation to follow the effects of diet and hormone levels on bone resorption over a woman's life. Results of biomedical significance are not yet available.

Dating cells

Atmospheric nuclear weapons tests in the 1950s and 1960s injected a large amount of radiocarbon into the environment. During 1963–4, carbon-14 concentrations reached a level in the northern hemisphere that was 100% higher than that of the pre-nuclear era. After the signature of the Nuclear Test Ban Treaty in 1963, the concentration of radiocarbon has been decreasing with a half-life of 15 years, due to the exchange of carbon with the biosphere and the oceans. In a few years, environmental radiocarbon will return to pre-nuclear era levels. The features of this 'bomb-pulse' have been determined by reading the record of carbon-14 in the atmosphere, tree rings, sediments, and ice cores. This effect is somewhat reduced in the southern hemisphere.

Scientists use the radiocarbon bomb pulse to measure cell turnover rates on the timescale of years and decades in organisms that have lived after the 1950s. The chromosomes in cells incorporate carbon-14 from the environment, and the concentration of carbon-14 in a living cell corresponds to the level of carbon-14 in the atmosphere at the time it formed. The genomic DNA, which remains stable after the last mitosis, is used for AMS radiocarbon analysis.

The dating of brain cells through the carbon-14 bomb spike has shown that neurons are not generated after birth in the human cerebral neocortex, where language and intelligence is located. This conclusion is valid within the detection limits of the method, which is based on the analysis of the DNA extracted from 15 million cells, necessary to obtain 30 micrograms carbon for the AMS measurement at better than 1% precision. The method has the potential to map cell renewal for the whole human body and answer questions about the response of critical organs to ageing.

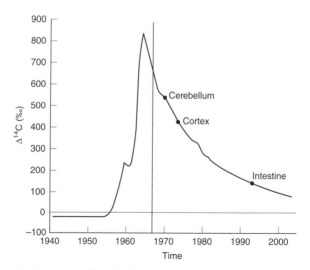

21. Dating human cells with the radiocarbon bomb pulse.
Radiocarbon levels (in relation to a universal standard, in per mil) in genomic DNA from the brain and the intestine of an individual born around 1967

The origin of certain pathologies can be verified, including the connection between lack of neurogenesis in the hippocampus and Alzheimer's disease, cell turnover in the eye's lens and cataracts, development of fibrotic material in the heart, and loss of cardiac function.

Chapter 5
Radioactive gadgets and gauges

In industry and at home

Industry has been using radioactive tracers and radiation-based devices to solve a myriad of problems for more than 60 years. A radionuclide tracer works by mimicking the substance that it is tracing, so that it tracks the chemical reactions and physical processes. For example, ^3HHO is used to trace water and $^{14}CO_2$ carbon dioxide. The principle of the radiotracer methodology consists in the injection of a tracer at the inlet of the flow system to be studied, followed by a measurement of the concentration at the outlet, at different times. Sealed radioactive sources are used in consumer products and in materials processing, for example to measure thickness, moisture, and other parameters.

Both radioactivity and radiation-based technologies have applications in mining, petroleum, energy, chemicals, paper, cement, and electronics, as well as in the automotive and aerospace sectors.

In the petroleum industry, advanced and interdisciplinary methods are increasingly used to study oil reservoirs. Difficulties in recovering the decreasing amount of remaining oil make it imperative to use all available tools, starting from radiotracers. Reservoirs are very complex systems, with different fluids flowing in heterogeneous porous materials, so the analytical models developed to describe their behaviour need to be validated with the direct measurement of key parameters, such as flow velocity.

Another special radiotracer application is related to metal release into the environment by solid waste incinerators. The ashes produced during this process contain high levels of copper and zinc, which could create problems for the environment. Gamma-ray emitters copper-64 and zinc-69(m) have been used in pilot plants to study the evaporation of copper and zinc, helping to identify the operating parameters of the plant that would minimize pollutant release.

Waste water treatment plants involve complex processes influenced by fluid flows. Optimization of the procedure is often carried out using radiotracers. Small concentrations of bromine-82 or tecnecium-99(m) are injected into the waste-water stream so they can help to estimate important parameters of the process to improve the waste removal efficiency.

Radiotracer imaging in 3-D is also being developed to study the flow of systems with many phases. SPECT and other techniques used in medicine are being considered, but the costs involved would presently limit their routine application.

Devices based on sealed radiation sources are used to monitor and control the thickness of paper, metal, and plastic foils. The target material is moved between a radioactive source, usually a beta source, and a detector that continuously measures the flux of particles. Thallium-204 ($T_{1/2}$ = 3.77 years) is commonly used in the paper industry. It emits beta particles of relatively high energy – 764 keV. Changes in thickness of the foil alter the flux of particles, hence the amplitude of the electronic signal from the detector. A computer monitors the detector and automatically adjusts the pressure and distance between the rollers. This method has the advantage of being low maintenance and non-invasive.

Gauging devices using neutron sources can be applied to the measurement of the water content of soil. Neutrons react very strongly with light elements such as the hydrogen in water, and their flux through the soil is related to its moisture. They are also used in road construction to gauge the density of road surfaces.

Methods routinely used in medicine, such as X-ray radiography and CAT, are increasingly used in industrial applications, particularly in non-destructive testing of containers, pipes, and walls, to locate defects in welds and other critical parts of the structure. Gamma sources, such as iridium-192, selenium-75, and ytterbium-169, are also used. Neutron radiography is finding new applications in the radiography of large metallic components such as the turbine blades of aeroplanes, where it can complement information obtained with other methods. For example, the use of high-energy X-rays, needed to penetrate thick components, give very poor contrast for the thinner parts.

Neutron tomography is also being developed, thanks to the progress in computer performance, neutron digital imaging, and neutron production, but the high costs hamper its commercial-scale use.

Meanwhile, ionizing radiation, usually gamma rays, is used to sterilize consumer products or medical products. As we mentioned in Chapter 3, food irradiation is another application, often using cobalt-60 sources. It destroys parasites and bacteria, extending the shelf-life of many products.

The manufacture of many modern products is based on the use of ionizing radiation. Non-stick cooking pans are irradiated to fix the coating to the base. Cosmetics for the skin and hair, and solutions for contact lenses are sterilized with gamma rays to eliminate allergens or irritants. Precious stones, including diamonds and amethysts, are irradiated with accelerators or reactors to change their colours.

Radioactivity or radiation-based consumer products are ubiquitous. Photocopying machines use small sources of radiation to remove static, whereas smoke detectors use alpha-emitters such as americium-241 (typically 0.9 microcuries). A small ionization chamber continuously detects the alpha particles, and smoke between the source and the detector reduces the amplitude of the detector signal, triggering the alarm.

During the first decades of the 20th century, radioactivity - based products proliferated, not always in useful applications. The Radium Corporation in New Jersey developed a big business using paints made of a mixture of radium, water, and glue for dials of watches and other instruments that needed to be seen in the dark. A special application was the use of radioactive paints in the dials of instruments for aeroplanes during the First World War. The young women doing this work licked the brushes, and many died as a result of chronic exposure to radioactivity.

Many other radioactive products appeared on the market in the good old days when a naive public was convinced that radioactivity would not only unleash unlimited energy, but that it also had curative and beautifying properties. Radithor was a well-known tonic made of distilled water containing radium, manufactured by William Bailey Radium Laboratories in the USA. Bailey also sold the Radiendocrinator, an instrument made of paper impregnated with radium to place under the scrotum at night to enhance virility. Other products enriched in the 'benefic' properties of radium and other radionuclides were the Tho-Radia face cream in France, the Doramad toothpaste, and the Radium Chocolate sold during the Second World War in Germany. Vita Radium suppositories were sold by the Home Products Company of Denver, Colorado. Radium water from the mines of Joachimstal was used to make bread in the Hippman-Black bakery in Bohemia. The Radiumchema Company in Joachimstal

22. A beauty cream containing thorium and radium, supposed to have curative and beautifying effects. It became popular in France during the 1930s

manufactured a radium source designed to add radon directly into a drinking-water glass, to take in case of headaches. Atomic Energy Lab toys sold in the early 1950s in the USA also contained small amounts of radioactivity.

During recent decades, the radioactivity passion ended, and all non-essential radiation-based consumer products were cleared from the market as measures to protect people and their environment were strengthened.

In space

In space exploration, instruments based on radionuclides are used to propel spacecraft and to generate heat and electricity, especially in deep space, where temperatures get down to around -270 °C, nearly absolute zero.

Radioisotope heater units (RHU) based on plutonium-238 have been developed, each weighing less than 50 grams and with dimensions of 2–3 centimetres. Alpha particles emitted by the radionuclide stop in a ceramic material, producing heat. Each unit produces only one watt, so several units must be used to be practical. RHUs are used in spacecraft to keep the instrument warm enough to operate, when solar energy is not available.

Electrical power is essential for the numerous computerized instruments used in flight control and data transmission. An increasing number of analytical systems are loaded in spacecraft to be used for *in situ* analysis on other planets, including miniaturized analytical systems based on radioactive devices and centimetre-sized spectrometers (like the one that will be sent to Europa, the icy moon of Jupiter, to analyse sulphur isotopes *in situ*, in the search for signatures of life). Radioisotope thermoelectric generators (RTG) are the devices of choice to provide the needed electricity, particularly when the aircraft is very far from the Sun.

RTGs are based on the use of plutonium-238, but in this case heat is converted into electricity using the Seebeck effect, in which a temperature gradient is transformed in a difference in voltage. Packages of several RTGs can generate tens of kilowatts without the need of movable components. These systems have been used in the Apollo, Voyager, Pioneer, Galileo, and Ulysses missions. Galileo, in its eight-year trip to Jupiter, used 120 radioisotope heater units and 2 radioisotope thermoelectric generators. *Curiosity*, a 900-kilogram rover equipped with advanced scientific instruments being sent by NASA to Mars, will explore the red planet using RTGs fuelled by the radioactive decay of plutonium-238.

In the past

The analysis of trace elements with neutron activation analysis, proton- and X-ray-induced fluorescence, and other nuclear techniques can reveal the source of materials in archaeological artefacts, ranging from obsidian tools to pottery and marble objects.

Neutron activation analysis, also developed by George de Hevesy, in 1936, is based on the detection of the radioactivity induced by the exposure to a neutron source. The emitted gamma rays are characteristic of the elements to be analysed. In fluorescence methods, protons or X-rays knock inner-shell electrons from the atoms. More external electrons then decay to the lower energy levels, emitting characteristic X-rays that can reveal the corresponding elements.

These methods are generally non-destructive, thus meeting one of the fundamental requirements for archaeologists and museum curators.

The nuclear physics laboratories located in Florence and at the Louvre in Paris are both dedicated to the analysis of paintings and other cultural heritage materials.

For example, nuclear techniques have been used at the Louvre to shed light on the 1521 art book of the German painter Albrecht

Dürer. Trace elements detected with proton-induced fluorescence revealed the genesis of his creation, including the origin of the materials used and the chronology of the works.

In Florence, scientists have analysed ancient manuscripts, fingerprinting the ink on the basis of its copper, zinc, and lead content. They were also able to date Galileo Galilei's notes on his discovery of the laws of motion, which was especially useful since the great Italian scientist had written on folios without a date.

Absolute dating methods used to pin down key turning points in our species' past are based on time-dependent effects related to natural radioactivity. The *in situ* production of long-lived radionuclides such as ^{10}Be and ^{26}Al through the bombardment of stone by secondary cosmic rays is used to date rock surfaces and artefacts. Thermoluminescence (TL), optically stimulated luminescence (OSL), electron spin resonance (ESR), and fission track dating all exploit the build-up of radiation effects in specific crystals. Finally, the build-up of a radiogenic daughter product from a primordial radionuclide forms the basis of potassium-argon, argon-argon, and uranium-series dating.

Carbon-14 is the chronometer most widely used in archaeology and environmental sciences. The American physical chemist and Manhattan Project veteran Willard Libby developed radiocarbon dating in 1946, and won the Nobel Prize for 'his method to use carbon-14 for age determination in archaeology, geology, geophysics, and other branches of science'. Libby's work started with the discovery of this rare radionuclide in methane from Chicago's sewage, adapting the isotope-enrichment techniques used in his weapons programme to his radiocarbon-dating research.

Carbon-14 is produced in the stratosphere by the nuclear reaction of cosmic neutrons with nitrogen:

$$n + {}^{14}_{7}N_7 \rightarrow {}^{14}_{6}C_8 + {}^{1}_{1}H_0$$

The radionuclide is then oxidized and mixed with stable atmospheric carbon dioxide, which is in equilibrium with the biosphere. Plants and animals absorb carbon from the atmosphere via photosynthesis and metabolic processes. Their carbon isotope ratios are close to those in the atmosphere. When carbon dioxide is taken up and fixed in organic compounds by photosynthesis, it is isolated from the atmosphere: the stable isotopes of carbon, carbon-12 and carbon-13, maintain their concentrations while carbon-14 decays, without being replenished, via the process:

$$^{14}_{6}C_8 \rightarrow {}^{14}_{7}N_7 + e^- + v^-$$

Libby evaluated the age of an organic sample by measuring the carbon-14 residual activity with an ionization chamber. The radioactivity was very small and the secondary cosmic rays further complicated the measurement. With his collaborators, he constructed a sophisticated system that switched off the central measuring chamber when a secondary muon reached the detector. This increased substantially the sensitivity of the system.

The next revolution in radiocarbon dating was the development of direct atom counting by AMS, already introduced in Chapter 3. In modern AMS spectrometers, specimens with just a few micrograms of carbon can be analysed in less than one hour. Samples of one milligram can provide ages for organic materials older than 50,000 years.

Radiocarbon dating revolutionized archaeology, providing a precise and direct measurement of the timescale for the late Quaternary. It revolutionized the understanding of European prehistory, previously dated only by correlation with the historical chronology of the Near East. Radiocarbon shed light on other archaeological mysteries, including the age of the Turin Shroud,

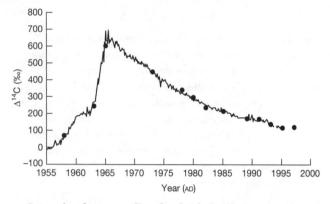

23. Comparison between radiocarbon levels (in relation to a universal standard, in per mil) in Australian wines (dots) and its concentration in the atmosphere of the southern hemisphere during the 'nuclear era' (line)

the end of the Mycenaean civilization, the Ice Man, the extinction of the Neanderthals and the arrival of humans in Australia, topics that we will discuss further in Chapter 8.

The radiocarbon bomb pulse, which was discussed in Chapter 4, offers another method for high-precision dating of organic materials during the last 60 years, providing chronological information with error of less than one year. In forensic science, it has been applied to evaluate the time of death of individuals. The most suitable materials are those having a fast turnover time of carbon, such as lipids from bone, bone marrow, and hair.

The carbon-14 bomb pulse has also been used for dating opium poppies, coca bush, and other illicit drugs. This information would support the action of law-enforcement authorities against criminal organizations involved in drug trafficking. It can also be used to determine wine vintages accurately, and to reveal the addition of unrelated materials of natural and synthetic origin.

Chapter 6
Fear of radioactivity

In 2007, the security officers at the checkpoints of the XV Pan American Games in Rio de Janeiro had mysterious gadgets hanging from their belts: personal radiation detectors. These highly sensitive radiation monitors were used as a first line of defence against radiation. The second line of defence was based on a more sophisticated gadget: the radionuclide identification device. This operation had to be carried out by radiation protection professionals with a good knowledge of radioactivity, isotopes, gamma-ray spectra, and germanium detectors. Whenever an alarm on a personal radiation detector went off, the people in the queue would be escorted by the National Security Force and scanned with the radionuclide identification device. The investigation would then move to the third line of defence, if any of the following cases was verified:

(a) the radionuclide was not used for medical treatments;
(b) the radionuclide identification device signalled the presence of neutrons, a signature of nuclear material;
(c) the radiation dose rate was larger than 100 mSv per hour (this limit had been set by the National Nuclear Authority).

The person, their car, or the radioactive container would be isolated for further investigation by a field response team, made up of experts in radiation detection and protection, environmental assessment, and by internal and external dosimetry. The team was equipped to analyse gamma rays, alpha particles, neutrons, and other ionizing radiation.

Other procedures were established to prevent criminal acts involving radioactive material. All open areas surrounding the venues were surveyed using car-borne gamma-ray-mapping equipment. The soccer field was surveyed by teams of analysts carrying gamma- and neutron-detection equipment on their backs, capable of radionuclide identification, along with global positioning systems.

All venues were surveyed, before the Games started, to measure the variability of the natural level of radiation, determining the baseline background at selected locations. Concrete, for example, has much higher baseline radiation levels than other building materials.

The Rio de Janeiro region hosts most of Brazil's nuclear installations: two nuclear power plants, a uranium-enrichment facility, four nuclear research reactors, and six nuclear research institutions; it also has 70% of the sealed radiation sources used in medicine, industry, and research in the country. These facilities were asked to strengthen their security, since they constituted a possible source of radioactive and nuclear material for nuclear terrorists. Patients using nuclear medicine were advised to carry a certificate stating the radionuclides they were using and their nuclear activity.

The personal radiation detector alarm was activated 42 times during the Games, 40 times by patients who had recently undergone nuclear medicine examinations, and twice by false alarms. The field response team was involved only three times, but eventually all of the alerts were false alarms, in terms of public threats.

Similar security systems were established with the help of the International Atomic Energy Agency for major public events, including the 2008 Olympic Games in China and the 2010 Soccer World Cup in South Africa. The fight against radiological terrorism is now protocol for national security forces dealing with events attracting large crowds.

Pierre Curie's reference to the dark side of radioactivity in his Nobel Prize ceremony was prophetic. More than a century after its discovery, radioactivity still arouses fear and suspicion.

The nuclear dilemma

At 8:15 am on 6 August 1945, the start of the nuclear era was heralded with the dropping of 'Little Boy' on Hiroshima. The uranium-235 bomb, with a power corresponding to 20,000 tonnes of TNT, killed 70,000 people instantly, while a further 70,000 died from the effects of radiation within five years. Three days later, the plutonium bomb nicknamed 'Fat Man' would be exploded over Nagasaki, killing a similar number of people. The previous month, during the first successful nuclear explosion at Alamogordo in New Mexico, the director of the Manhattan Project, Robert Oppenheimer, had said: 'I have become Death, the destroyer of the worlds.' After the explosions on Hiroshima and Nagasaki, he admitted that 'the physicists have known sin; and this is a knowledge they cannot lose'.

Since then, the mushroom clouds of Hiroshima and Nagasaki have cast a dark shadow over any human endeavour related to radioactivity.

In the post-war years, other countries developed nuclear capability. The USSR tested its first plutonium bomb in 1949 and its first hydrogen bomb in 1953. Other countries soon joined what is informally known as the 'Nuclear Club': the United Kingdom in 1952, France in 1960, China in 1964, India in 1974, and others later on.

At the same time, other sectors of society were promoting civil applications of nuclear science with the potential of deeply transforming our lives.

Lewis Strauss, the first chairman of the US Atomic Energy Commission, is quoted as saying, in a 1954 speech to the National Association of Science Writers, that 'our children will enjoy in their homes electrical energy too cheap to meter'. Hopes were high that the nucleus would power our homes, our industries, and even our cars.

American comic books during the nuclear era reflected the duality of radioactivity. In the fictitious world of superheroes, radiation could either be harmful or beneficial: kryptonite radiation could kill Superman, but on the other hand, it was the bite of a radioactive spider that gave Spiderman his powers. Gamma radiation transformed Bruce Banner into the monstrous Hulk; however, the Fantastic Four gained their superpowers after being exposed to cosmic rays in space. Comic-book writers seem to have been even more obsessed by radioactivity than Marie Curie!

Although nuclear science and technology did not live up to all its expectations, the post-war years did see the development of many applications in diverse sectors, including industry, medicine, and agriculture. On the fiftieth anniversary of the IAEA, the agency's assistant director general, David Waller, said:

> ...the pressing issue was how to further develop and promote these peaceful applications, while at the same time prevent the spread of weapons technology. That was – and indeed still is – the nuclear dilemma.

This attitude had been clear from the very start. In 1953, US President Dwight Eisenhower proposed to the United Nations a programme called 'Atoms for Peace', and promoted the creation of the IAEA with the two-pronged mission to promote peaceful nuclear applications of high socioeconomical significance while arresting the arms race.

During many years of diplomatic negotiations, the international community built a new legal framework to constrain the development of nuclear weapons. The Treaty on the Non-Proliferation of Nuclear Weapons (also known as the Non-Proliferation Treaty, or NPT) was signed in 1970. Under this treaty, nations could develop peaceful nuclear applications but had to cease building nuclear weapons. The main aim of the then nuclear states, the USA, USSR, China, France, and the UK, was to preserve the *status quo*, confining nuclear weapons to their small group without any serious attempt to proceed towards their 'denuclearization'. The NPT succeeded only partially, and the nuclear club expanded. India, Pakistan, and Israel did not sign the NPT, while the Democratic People's Republic of Korea signed but withdrew in 2003. Other countries, such as Libya and Iraq, in spite of having signed the NPT, considered developing nuclear weapons programmes, but eventually decided otherwise. The discussion on nuclear programmes in Iran is still open. South Africa is the only country that, towards the end of the apartheid regime, voluntarily gave up its nuclear arsenal of six nuclear weapons that it had assembled between 1960 and 1980.

Between the early 1960s and the beginning of the 1990s, the international safeguards system implemented by the IAEA was developed to provide assurance that nuclear materials had not been diverted from declared nuclear activities. These traditional safeguarding methods are based on nuclear accountancy, complemented by containment and surveillance techniques. Since the early 1990s, work has been under way to strengthen the international safeguards system by giving it a capability to verify the absence of undeclared nuclear material and activities. The IAEA has assumed the role of a detective, with the task of verifying the correctness and completeness of declarations made by countries adhering to the safeguards agreement.

As part of the strengthened safeguards system, the IAEA has implemented environmental sampling and analysis. Highly enriched uranium (or plutonium) is needed for nuclear weapons, therefore abnormal uranium isotopic ratios are the clearest signatures of enrichment. The reprocessing of irradiated reactor fuel for the production of plutonium releases fission products and actinide isotopes into the surrounding environment. Iodine-129 and uranium-236, both measurable by AMS in environmental samples, are key radionuclides for identifying clandestine nuclear activities.

New tools for the nuclear watchdogs

Antineutrinos are continuously emitted by neutron-rich fragments produced by the fission of uranium and plutonium nuclides in nuclear reactors. The fuel rods contain both uranium-238 and uranium-235, with the latter being the major fissile nuclide in existing reactors. A fraction of uranium-238 nuclides absorbs neutrons and then decays into plutonium-239, which also fissions, producing fragments that emit antineutrinos.

Antineutrinos from reactors were first detected 50 years ago. It was later suggested that they could be used in nuclear safeguards to detect clandestine diversions of nuclear material from nuclear power plants into nuclear weapons programmes. The IAEA applies practices of containment and surveillance, but these are very expensive and time-consuming. It would be much more effective to measure the inventory of the nuclear material directly while the reactor is operating, which is what antineutrinos can do.

The antineutrinos count rate is proportional to the reactor thermal power and to the reactor fuel composition. The first correlation derives from the proportionality of the antineutrino count rate to the fission rate of uranium-235. The second correlation derives from the different antineutrino energy spectrum for uranium-235 and plutonium-239. The average number of antineutrinos per fission released by uranium-235 and plutonium-239 is 1.92 and 1.45, respectively. The antineutrino detection rate decreases at a known rate during a reactor fuel cycle as uranium-235 decreases and plutonium-239 increases, stepping up again to the original value after the reactor has been refuelled with 'clean' uranium.

One antineutrino detector, being tested in the USA by the Sandia National Laboratory and the Lawrence Livermore National Laboratory, is a liquid scintillator doped with gadolinium.

The collision of an antineutrino with a proton of the scintillator produces a positron and a neutron. The positron immediately generates a flash of light in the scintillator and then annihilates an electron, producing two gamma rays that generate further flashes. After a 30-millionths of a second delay, the neutron is captured by the gadolinium nucleus, producing more gamma rays, which in turn produce other flashes in the scintillator. The flashes then become electronic signals that can be stored in a computer and analysed. The characteristic sequence of signals is used to count the relatively rare antineutrinos from the fissile material that interact with the detector, discriminating against the background noise produced by other particles and radiations.

The detector, located 25 metres from a PWR reactor, would detect 400 antineutrinos per day and track with high precision the history of the reactor's operations, including emergency shutdowns and refuelling, and check the inventory of fissile material.

Nuclear terror

In the past 20 years, a new political landscape has taken shape, with ethnic and religious tensions exacerbating global socioeconomic problems, causing conflicts in several global hotspots. The fear built on the increased alert developed during the 1990s after the collapse of the USSR, one of the major nuclear powers. After the attack on the Twin Towers in 2001, the spectre of nuclear terrorism and 'dirty bombs' strengthened public opposition to nuclear activity. The IAEA had to develop a strong security programme with urgency, to protect nuclear materials and facilities from potential terrorist attacks, and to prevent the malicious use of radioactivity.

Four nuclear terrorism scenarios have been considered by the IAEA: (1) the theft and detonation of an existing nuclear weapon; (2) the theft or purchase of fissile material to manufacture and detonate an improvised nuclear device; (3) attacks on nuclear facilities leading to the release of radioactive material; and (4) the illegal acquisition of radioactive materials for the manufacture of a 'radiological dispersal device' or a 'radiation emission device'.

The construction of a nuclear device would require considerable infrastructure, expertise, and money. It is generally accepted that the most likely method by which terrorists could obtain nuclear material and technical information would be through existing state-owned facilities. For this reason, the proliferation of nuclear materials, particularly in the past 20 years, has raised fears in the international community as it increases the probability of such materials falling into the wrong hands.

Terrorists could expose the public to radiation and contaminate the environment by attacking a nuclear facility. However, all states require strict security for such facilities, particularly for nuclear power plants.

What are the potential sources of nuclear and radioactive materials, which could be used by terrorists? How many nuclear facilities could be their target?

There are more than 10,000 cobalt-60 ($T_{1/2}$ = 5.3 years) and caesium-137 sources used in hospitals for teletherapy, mostly in developing countries, with intensities of tens of thousands GBq (one gigabecquerel equals one billion becquerel). In total, there are more than 100,000 radioactive sources in the world, mostly used in hospitals and industry, that are categorized as very dangerous by the IAEA. There are more than one million weaker radioactive sources. We should also add to these the over 25,000 nuclear weapons, some 3,000 tonnes of highly enriched uranium and plutonium, and more than 1,000 nuclear centres with power reactors, research reactors, and centres for nuclear material processing and storage.

The use of radiological dispersal devices, or radiation-emission devices, is considered as the more likely scenario, since radionuclides are widely used in medicine, industry, and science, and therefore are potentially accessible to criminals and terrorists.

Radiological dispersal devices could use conventional explosives to spread radioactive material. If detonated in a city, they would cause mass hysteria and force the evacuation of the affected area, during a lengthy and costly clean-up.

A radiation emission device would consist of a radioactive source planted in a location where it could dose a target undetected for a long time.

While the use of dispersal or emission devices would be unlikely to kill many people, they would wreak havoc, and have thus been labelled 'weapons of mass disruption'.

Protecting nuclear and other radioactive materials is a major concern of the international community, and there have been considerable efforts to modernize physical protection and accounting systems throughout the world. States and international organizations have also been providing technical and financial support to developing countries. The G8 group of nations is supporting former Soviet Union states to manage and secure their radioactive materials. RTGs using strontium-90

($T_{1/2}$ = 28.8 years) sources were heavily used in the USSR to provide electricity and heat in remote locations, particularly in military plants. A typical source would emit beta rays corresponding to millions of GBq, providing several kilowatts of thermal energy that could be converted into electric energy. Hundreds of RTGs were used to power lighthouses along the Siberian coasts. Some RTGs are still active, but they are being replaced by systems based on solar or wind energy.

In the past, before the 2001 terrorist attack on the World Trade Center, security measures applied to non-fissile radioactive materials were focused on preventing accidental access or petty theft. Few took the threat of radiologic terrorism seriously, and systems were poorly regulated. Many states are now addressing the problem, but there are still thousands of unaccounted radioactive sources worldwide. These 'orphaned radioactive sources' are operating outside official regulatory control or are hidden. They may have been lost, discarded, or stolen, and thus constitute potential weapons for terrorists.

The IAEA has promoted nuclear forensics laboratories specializing in the analysis of smuggled radioactive materials seized by customs and other organizations, to obtain clues on the origin of the material. Sensitive techniques such as AMS are used to determine the isotopic ratios of nuclear materials, such as plutonium and uranium, at very low concentrations. These ratios are sensitive indicators of the past history of the material. In particular, the isotope uranium-236 is useful for reconstructing the irradiation history of uranium samples.

Fortunately, radioactivity also has a bright side. In fact, it has the power to shed light on the origin of the universe and terrestrial life, including our own. Physicists have been intensively involved as a result in a very wide spectrum of studies, in collaboration with a number of scholars in other disciplines.

Chapter 7
Tracing the origin and the evolution of Earth

The X-ray beam of the Swiss Light Source, with an intensity millions of times stronger than that used by Röntgen to irradiate the hand of his wife, has been used to take images of a much older form of life: multicellular organisms found in south-eastern Gabon that had been alive 2.1 billion years ago.

The centimetre-sized fossilized structures from that geologic formation have been analysed, by taking thousands of digital X-ray radiographies from different angles using an advanced detector that combines a scintillator and a CCD camera. These powerful X-rays are generated from electrons rotating at nearly the speed of light at high vacuum in a doughnut-shaped storage ring with a circumference of 288 metres: the heart of the synchrotron radiation facility at the Paul Sherrer Institute in Switzerland. Synchrotron radiation is generated when the electron's trajectory is deviated by magnetic fields, an effect known since 1947. X-rays of the desired energy are selected by a monocromator, penetrate the sample, and are finally detected by the scintillator, which converts them into visible light. Projection images are magnified by microscope optics and digitized. Powerful algorithms reconstruct the microtomographies, showing intimate morphological details in 3-D, with a resolution thousands of times better than in hospital CT scans. This analysis has confirmed that the fossils embedded in the rock were authentic multicellular life forms striving to survive during the Paleoproterozoic, some 2.5–1.6 billion years ago.

The ages of the African rocks containing the small organisms have been determined using uranium radioactivity. Uranium is present in small concentrations in the microscopic zircon crystals that took form when the rock solidified. Uranium-238 decays through a series of radioactive steps, ending with a stable nuclide, lead-206. In modern analytical systems, lead-206 is measured with high precision on an ion microprobe. A beam of charged particles drills micron-sized holes in the zircon grain blasting out its atoms. The atoms are sent to a mass spectrometer, which sorts them according their masses and counts the concentration of lead-206 atoms. The relative concentration of uranium-238 and lead-206 reveals the sample's age.

The origin of multicellular life is pivotal to understanding evolution. In the ancient black shale from Gabon, Röntgen and Curie meet Darwin, who died only a few years before they discovered X-rays and radioactivity. One hundred and fifty years later, X-rays and radioactivity form the basis of the instruments that provide critical information supporting Darwin's ideas.

What follows is an outline of some of the key chapters in the history of the Earth from its genesis, where radioactivity and radiation play a critical role.

Genesis

One-trillionth of a second after the Big Bang, around 13.72 billion years ago, with temperatures soaring at 1 quadrillion K, the universe was a homogeneous mixture of quarks, electrons, neutrinos, and other elementary particles (and the corresponding antiparticles). High-energy accelerators can produce these particles by colliding head-on beams of electrons, protons, or other ions, accelerated to velocities approaching that of light

CERN's Large Hadron Collider can presently collide protons against protons at an energy (in the centre of mass) of 7 TeV (trillion electron volts). The LHC will eventually collide them at energies of 14 TeV, recreating in the laboratory the conditions of the universe just after the Big Bang.

The first isolated neutrons and protons, the building blocks of all nuclides, formed from the quarks that survived the annhilation with anti quarks between one microsecond and one second after the Big Bang, at 1 trillion K.

About 3 minutes after the Big Bang, with the temperature below one billion K, the expanding universe started functioning as a nuclear fusion reactor, forming deuterium, tritium, helium and other light nuclides (this is the so-called 'Big Bang nucleosynthesis').

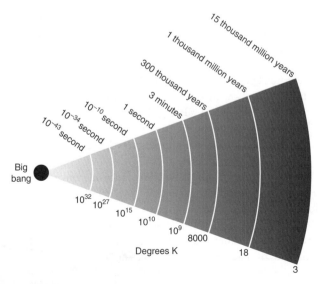

24. History of the Universe

The universe continued expanding and cooling, and nuclear fusion reactions stopped about 20 minutes after the Big Bang, when temperatures were about 300 million K. At this stage, the composition of the universe, not considering dark matter, was approximately 75% hydrogen-1, 25% helium-4, 0.01% hydrogen-2, with some traces of helium-3, lithium-6, lithium-7, beryllium-9, boron-10, and boron-11. Radioactive nuclides such as hydrogen-3, beryllium-7, and beryllium-10 were also originally present but have all decayed, so far.

About 377,000 years after the Big Bang, when the temperature dropped below 3,000 K, the electromagnetic force started binding electrons to existing nuclei to form the atoms of hydrogen and helium, plus other light elements produced during primary nucleosynthesis. Photons were not scattered by free electrons any more, so they could finally travel through the neutral atoms of the universe. The physicists Arno Penzias and Robert Wilson discovered this primordial radiation in 1964, producing the most persuasive evidence supporting Big Bang cosmologies. They received the Nobel Prize in Physics in 1978 'for their discovery of cosmic microwave background radiation'.

After a quiet period of about 150 million years, the so-called Dark Era, the force of gravity started condensing hydrogen and helium into large gas clouds and galaxies, similar to our own Milky Way. Two billion years after the Big Bang, the universe looked like an immense desert punctuated with clusters of galaxies.

Within each galaxy, matter continued condensing under the action of gravity, and the temperature started increasing locally, forming the first generation of stars.

After the temperature in the core of the stars soared above 10 million K, the nuclear force pulled nucleons together to form helium, as in the first phase after the Big Bang. Nuclear fusion released large amounts of energy, 'switching on' the stars and making them shine.

Once the core of a star started running out of hydrogen, pressure from fusion reactions diminished, gravity became the dominant force, and the core of the star collapsed, while the outer part expanded, transforming the star into a *red giant*.

For more massive stars, the stellar evolution continued. After the temperature of the stellar core reached 100 million K, helium's fusion reaction kicked in, forming heavier nuclei. When the helium supply became exhausted, the core was made of carbon and oxygen.

Gravitation took over again increasing the core density and, when the temperature reached 1 billion K, first carbon-12 and later oxygen-16 underwent fusion reactions forming heavier nuclides, including neon-20, sodium-23, silicon-28, and phosphorus-31. When the core temperature went up to 4 billion K, silicon-28 burned via a sequence of alpha-particle reactions, producing the elements of the iron group.

The sequence of these 'hydrostatic fusions' ended at the iron stage. The 'stellar onion' (with an iron core) contracted without triggering additional fusions and finally exploded, becoming a *supernova*. The energy of the shockwave produced elements heavier than iron in shells around the core of the star through the capture of neutrons, mainly originated from alpha-induced reactions on carbon and neon. The slow neutron-capture mechanism, the so-called *s*-process, in which neutrons are captured at intervals of more than 10 years on average, can produce nuclides up to bismuth-209. The rapid neutron capture, the *r*-process, in which neutrons are captured at intervals of less than 1 second, can produce heavier nuclides, up to thorium-232, uranium-235, uranium-238, and plutonium-244. The production ratios of thorium-232 to uranium-238 and of uranium-238 to uranium-235 indicate, assuming a constant nucleosynthesis rate, that our galaxy has an age of 12.8 ± 3.0 billion years.

The supernova explosion ejects a variety of nuclides into space, accelerated by supernova remnants at speeds up to 100 kilometres per second. The only leftover after the explosion is a neutron star or a black hole, depending on the mass of the star.

Supernova explosions are the final destiny of very large stars, which shine only for a few billion years, while smaller stars, like our Sun or smaller ones, will end their fusion process less violently, with a life of more than 10 billion years, becoming first *white dwarves* and finally *black dwarves*.

During the first half billion years of the Milky Way, massive stars evolved via the nuclear reactions that produced all the elements of our ordinary matter, including carbon, nitrogen, oxygen, silicon, magnesium, iron, through uranium, spreading them all over the galaxy by supernova explosions. They were included in the interstellar dust that compacted into the first solid bodies and became larger and larger via accretion. Our solar system formed a few billions years later from the ashes of the stars exploded in the early life of the galaxy. The atoms of our own bodies, which evolved to the present anatomical form 200,000 years ago, were once in the core of massive stars.

The high-energy protons and alpha particles produced by supernovae in our galaxy (and in other remote galaxies) are still now bombarding material in our solar system, including the Earth, while moving in a cold universe, at a temperature of 2.725 K.

Cosmic reactions

On 12 April 1912, a day for which a total solar eclipse had been forecast, Victor Hess, a 29-year-old Austrian physicist, loaded his electroscopes onto a balloon and rose into the skies of Vienna, up to more than 5,000 metres. He wanted to check whether the Earth's surface was bombarded by radiation of cosmic origin, a new idea of its time. After more than two decades of radioactivity studies, using improved detection systems, many scientists were convinced that the Earth's crust was not the only source of natural radiation. The previous year, a German scientist had taken an electroscope to the top of the Eiffel Tower, but the experiment had not been conclusive. Hess was convinced that this mysterious radiation was made of high-energy gamma rays that he might identify by moving further away from the Earth's surface. During his balloon ascent, the Austrian scientist noticed that the ionization due to radiation decreased up to 1,000 metres and then started increasing, eventually exceeding that on the ground. He found that radiation rates at 5,000 metres doubled those at sea level. He also noticed that the solar eclipse did not make any difference, proving that the highly energetic radiation was not coming from the Sun but from other sources in the galaxy, or from other galaxies.

Hess received the Nobel Prize in 1936 'for his discovery of cosmic radiation'. It has recently been reported that in 1911 an Italian physicist, Domenico Pacini, who was in contact with Hess, was the first to prove the existence of cosmic rays by exploring the sea rather than the sky. Pacini sank an electrometer sealed in a copper box to a depth of a few metres in the Gulf of Genoa and in the Lake of Bracciano, in Italy, measuring a strong decrease in radiation. This finding confirmed that the radiation was not from the terrestrial crust but from the cosmos. Pacini died in 1934, missing the chance of getting the Nobel Prize.

Victor Hess fled to the United States in 1938 after Hitler invaded Austria. At the beginning of the nuclear era, he became involved, again, in experiments on radioactivity from the sky, but this time 'man-made'. During the Cold War, he could be found on the top of the Empire State Building in New York testing radioactive fallout.

High-energy cosmic rays – mainly protons (~90%) and alpha particles (~9%), plus a minor component of electrons and gamma rays – interact with the nuclides of the elements that form the atmosphere, nitrogen (78%), oxygen (21%), argon (0.9%), plus other minor gases. The cosmic-ray flux depends strongly on energy, for example one particle per m^2 per second has an energy of about 100 MeV. There are particles with energies 100 million times larger than those produced at the Large Hadron Collider, but their fluxes are very small (less than 1 particle per km^2 per second).

The products of cosmic nuclear reactions with the atmosphere include long-lived radionuclides carbon-14, beryllium-10, aluminium-26, and chlorine-36. The cosmic-ray interaction with the atmosphere produces showers of secondary particles that reach the lithosphere, where they interact to produce additional radionuclides. Neutrons and muons are the main products from the primary interaction of cosmic rays in the upper atmosphere, present at the Earth's surface. Neutrons produce long-lived radionuclides in the first 3 metres of soil or rock. Muon capture and nuclear reactions of fast muons predominate at greater depths. Oxygen and silicon are the major target elements for producing beryllium-10 and aluminium-26, respectively. Chlorine-36 is produced by high-energy reactions on Ca and K and thermal neutron capture on chlorine-35. Neutron capture on calcium-40 produces calcium-41.

As we have seen in previous chapters, AMS is the technique of choice to analyse long-lived cosmogenic radionuclides with the required sensitivity.

A larger concentration of radionuclides can be produced by the interaction of high-energy cosmic rays on meteorites and other solid bodies in the solar system that are not shielded by an atmosphere.

Meteorites

In 1969, the same year that *Apollo-11* was aiming at the Moon to collect material to analyse back on Earth, two big chunks of space rocks hit our planet. The impacts provided scientists with precious extraterrestrial material. One meteorite crash-landed near Pueblito de Allende in Chihuahua, Mexico; another near the town of Murchison in Victoria, Australia. The Allende meteorite provided evidence on the early history of the solar system, while the analysis of the Murchison meteorite revealed protein amino acids and DNA components, supporting speculation that life on Earth had been seeded from another planet. Until the *Apollo-11* mission, meteorites were the only extraterrestrial material available to us.

Today, the relatively undifferentiated material composing meteorites continues to be a unique source of information on the materials of the solar system during the earliest phases of its history.

What scientists found in the Allende meteorite were small, millimetre-sized white and pink inclusions enriched in aluminium and calcium (known as CAI, calcium-aluminium-rich inclusions). Their findings are consistent with computer models of nebular cooling. The structures have high concentrations of uranium and can be dated with high precision using the uranium-lead method.

CAIs in meteorites like Allende were dated to 4.5672 ± 0.0006 billion years using uranium dating. This gives a high-precision timing for the creation of the first small solid pieces of material of the solar system from the protostellar nebula. Uranium dating provides a mean age of 4.550 ± 0.003 billion years for Earth's accretion, meaning that our planet formed rapidly in geological terms, less than twenty million years. During this period, all the small pieces of primordial solid material, such as CAIs, were orbiting the Sun and colliding. Some accreted, and big lumps got even larger. Eventually, this random process created the Sun, nine planets, tens of satellites, and thousands of asteroids (counting only those visible by telescope). The leftover material includes hundreds of thousands of comets, meteoroids, interplanetary dust, and plasma (the solar wind).

Many of the meteorites, such as Allende and Murchison, originated in the asteroid belt, between Mars and Jupiter. They were the primordial materials that could not be accreted into a tenth planet.

Metre-sized meteors travelling in the solar system are exposed to energetic particles, mainly protons, from the galaxy. Cosmic rays and their secondary products, including high-energy neutrons, interact with specific nuclides of the meteoroid, producing long-lived radioactive nuclides with half-lives suitable for studying their chronology. Beryllium-10 ($T_{1/2}$ = 1.38 million years) and aluminium-26 ($T_{1/2}$ = 0.70 million years) are probably the most common. Of course, short-lived radionuclides can be analysed in meteorites only if they are measured immediately after impact. Long-lived radioisotopes can be measured, instead, thousands of years after the strike.

Cosmogenic radionuclides start accumulating in the meteoroid after the rock is blasted from a large body, usually an asteroid, where they had been shielded from cosmic rays. The concentrations of cosmogenic nuclides depend on the irradiation geometry and on the duration of the cosmic ray irradiation. These factors are in turn determined by the collisional and fragmentation histories of the objects. Measurement of cosmogenic nuclides in extraterrestrial materials is but the first element in a chain of information that will eventually determine the irradiation history of meteorites in space. Cosmogenic radionuclides reveal that many chondrites – stony meteorites containing small inclusions called chondrules – have been exposed in space for millions of years.

Some meteorites collected from the blue-ice fields of Antarctica have been identified as being of lunar origin (ALHA 81005) or Martian (ALHA 84001). The analysis of beryllium-10 and aluminium-26 in ALHA 81005 showed that this object had spent less than 100,000 years in space, supporting the theory of a lunar origin.

Scientists worked out the transit time of ALHA 84001 from Mars to Earth by measuring the density of nuclear tracks produced in grains of pyroxene – silicate minerals common in meteorites and terrestrial basalts – from the rock. As with the Murchison meteorite, this meteorite was linked to possible extraterrestrial life.

Evolution of the solid Earth

According to Australian Aborigines, 'dreamtime' is the time of creation. In one of the dreamtime stories, told by the elders of the Gajerrong people of the Kimberleys, north-western Australia, a male belonging to a group of ancestral people, the Djibigun, desired a female named Jinmium. He chased her across deserts and swamps. When he finally reached her, she turned into a rock to escape him. Now this rock is a large sculpted boulder called Jinmium, the landmark of a well-known archaeological site. Science tells other stories of the creation of the solid Earth, often based on Australian rocks. A crucial role is still played by radioactivity-based methods.

When the Earth formed 4.55 billion years ago, condensing from the spinning solar nebula, it was a molten sphere made of iron-rich metal, silicate minerals, and volatile compounds. As material was collected to form the proto Earth, its temperature increased, partly due to the energy deriving from the gravitational accretion process and partly from the energy released by the radioactivity of uranium, thorium, and potassium. The heating created an unstable system, with lighter, hot material rising and being replaced by colder material that was sinking. During the first hundreds of thousands of years, a violent convection process was transferring heat from the layered core at 4,000–5,000 K to the mantle. Some crust started developing from the mantle, but it was recycled to the interior of the planet immediately after solidification.

Analyses of zircon minerals from Australia show that some of them derive from material that was part of the young Earth and that had not been recycled.

A zircon grain from Jack Hills in Western Australia (800 kilometres north of Perth) gave a uranium-lead age as old as 4.4 billion years, showing that, 150 million years after the birth of our planet, solid material was already appearing as granite rocks. The analysis of the crystal revealed that, in spite of the high temperatures, there was already water on Earth's surface during the geological aeon called Hadean ('time of Hell'). The temperature had decreased below the boiling point of water.

After a while, around 3.8 billion years ago, at the early stage of the Archaean aeon, cooling of the Earth slowed the convective movements, and the continental crust developed. Australian rocks show that the first cyanobacteria (cells without the nucleus, called prokaryotes) evolved around this time, probably near submarine hydrothermal vents. Subsequently, they formed mound-like structures (stromatolites) in the marine environment, as shown for some rocks from the Pilbara region in Western Australia which have been dated with uranium-lead to 3.5 billion years ago.

The oldest rocks from this period, which still exist today, are in the old Precambrian shields of Western Australia, South Africa, Greenland, and North America. These rocks are the result of repeated cycles of transformation that changed the structure and mineral composition of the original plutonic rocks, which formed more than 4 billion years ago. The transformations included high-temperature and high-pressure processes of younger rocks (known as 'metamorphism'). The processes outlined above led to the formation of the Earth structure that exists today – a solid inner core, liquid outer core, mantle, and crust.

By about 2.5 billion years ago, at the beginning of the Proterozoic aeon, the Earth's older crust, similar to the present lunar or meteoritic material, was replaced by a lighter silica-rich crust. In particular, rocks of this age show that this light crust is composed of discrete elements, called plates, which bear the continents. The plates stay upon the heavier iron- and magnesium-rich rocks of the upper mantle and move around, changing the arrangement of the lithosphere in a process called plate tectonics.

Cyanobacteria were producing significant levels of free oxygen. The presence of abundant oxygen released by the bacteria into the ocean water is testified, for example, by the banded iron formation found in the Precambrian rocks of the Pilbara region. These minerals are today a significant contribution to Australia's GDP. After 2.5 million years ago, oxygen started accumulating into the atmosphere. The presence of oxygen (including the ozone layer that blocked ultraviolet radiation) allowed the development of the first complex cells, eukaryotes, which probably evolved through endosymbiosis around 2.1 billion years ago.

Proterozoic rocks from the Flinders Ranges in South Australia show the glacial periods that occurred between 800 and 600 million years ago. The end of the Proterozoic is marked by the Ediacaran period (630–542 million years ago) when multicellular soft-bodied organisms evolved. The boundary between the Ediacaran and the following Cambrian period is also clearly marked in the rocks of the Flinders Ranges, with the appearance of sea creatures with shells, scales, and armour plating.

In the meanwhile, the geological evolution continues and is recorded in the terrestrial rocks. The basaltic magma comes up from the upper mantle through the fissures constituting the mid-ocean ridges, a process associated with plate divergence. The direction of the terrestrial magnetic field, resulting from the rotation of the outer core, is recorded in the magnetization acquired by the crystals of the basaltic materials as they cool. The Earth's magnetic field has changed in the last 3.2 billion years. Its excursions and reversals are reflected in the material extruded from the ocean floor. The absolute timescale can be reconstructed by using, again, radioactivity. In 1948, the American physicist Alfred Nier showed that potassium-40 decayed to argon-40. He argued that this was a perfect geochronometer to date the basalts in which the geomagnetic reversals were recorded. The half-life of the decay is 1.248 billion years, allowing to date all of Earth's history. The youngest ages for reversals are located near the ridges; the oldest, near the continents. The ages of the rocks on either side of the ridges are the same. This is evidence that the ocean floor opened up as the continents drifted apart.

When two plates converge, one might be forced beneath the other, triggering volcanism and carrying crustal material into the mantle. At continental margins, the subducted material essentially consists of oceanic crust and the sediment that covered it. Extensive geochemical work has shown that volcanism at continental margins returns some of the subducted material to the surface. Deep-sea sediments have a high concentration of cosmogenic radionuclide beryllium-10. Analysis of beryllium-10 in lava samples from island-arc volcanoes provided direct evidence for the incorporation in lava of oceanic sediments from subduction zones. The work has verified that oceanic sediments can be recycled and put a figure on the time of the process.

The cycle of exchange between the Earth's inner and outer layers has been going on since the so-called 'dark ages' of our planet. Crustal rocks are eroded, dumped into the ocean, and eventually pulled back into the mantle at subduction zones where continental and oceanic crusts grind over each other. The cycle starts at the lithosphere–asthenosphere boundary with molten mantle rocks, more than 100 kilometres underground.

The age of the Earth

The debate over the age of the Earth has been highly controversial. Until the 17th century, the discussion on the origins of the Earth and the universe had been mainly confined to priests, shamans, prophets, and philosophers.

According to the Bible, for example, the Earth was created on the evening of 22 October 4004 BC, as calculated by James Ussher, Archbishop of Armagh in Ireland, using the genealogy of Adam's family. Noah's flood had moved fish and other ocean animals to the top of mountains, where they can be found today in stratified rocks.

New ideas on this matter developed during the 18th century, but the information from the natural scientists of that period was much less 'precise' than that provided by Archbishop Ussher. In 1785, for example, James Hutton, the 'father of geological sciences', told the natural science scholars attending a meeting of the newly created Royal Society that our planet was shaped the way we see it today by slow geologic processes that acted over a very long period of time. The same environmental forces were acting in the present to shape the geological landscapes of tomorrow. Hutton concluded that in the geological history of our planet '…we find no vestige of a beginning, no prospect of an end'.

Geologists started looking systematically at fossils, correlating their order in the geologic sequence across the continents. Charles Lyell, a geologist, born the day Hutton died, evaluated the length of time of specific geological phenomena, such as the deposition of lava on Mount Etna, and then used the fossils to evaluate the entire geologic timescale of the planet. In 1833, he summarized all the available information in a book, *Principles of Geology*, which was an important source of inspiration for Charles Darwin, during his trip aboard the *Beagle*, as it seems to agree with Darwin's own ideas of the long time span needed for the evolution of life on Earth.

By Darwin's time, geologists had recognized and named on the geological column most of the sequences, characterized by their fossils. At the bottom was the Precambrian, with very little evidence of life. Then came the Cambrian, with an explosion of life forms, recognizable with the first biomineralized structures. Then followed the Ordovician, with its trilobites; the Silurian, with the first bony fishes; and after that, the Devonian, characterized by its red sands, fish, and earliest tetrapods. During this period, seeds evolved, allowing the definitive colonization of the land by plants. The Carboniferous appeared later on, with a highly oxygenated planet covered by lush vegetation that housed giant arthropods and the first small reptiles. When it disappeared, it gave way to the deserts of the Permian and the Triassic. At the end of the Permian, a period of dry and variable climate, most life on the planet, including trilobites and other marine life, along with reptiles and proto-mammals on land, suddenly became extinct. A warm tropical planet appeared again in the following period, the Jurassic, with the spreading of ammonites and other creatures of the sea and the domination of the dinosaurs. The next geological period, the Cretaceous, is mostly composed of chalk from an immense number of microscopic algae that covered the planet. It ends with the impact of an asteroid hitting the Earth. Then came the Tertiary, and finally the Quaternary – the 'dirt on top', according to the hard-rock geologists. The existence of a geochemically based new epoch termed 'Anthropocene' is currently a matter of discussion among the sedimentologists.

To measure the age of Earth, the geologists needed the physicists, with their quantitative methods, who could help them to conceive the 'clocks' needed to measure very slow geological processes, occurring over long periods of time.

Physicists, who were then called 'natural philosophers', entered the Earth dating game in the middle of the 19th century. Lord Kelvin, the father of the absolute temperature scale named after him, applied his knowledge of the laws of thermodynamics to the thorny problem of dating the Earth. Given that temperatures increase the more one proceeds towards the interior of the Earth, he assumed that our planet was cooling down from an original molten globe. Knowing the temperature at which the rocks melted and the cooling rate, he could evaluate the age of the solid Earth's crust. This was not a new idea. Isaac Newton had already speculated that it should have taken at least 50,000 years for the Earth to cool. In the 1760s, the French nobleman Georges-Louis Leclerc, comte de Buffon, made an experiment with hot iron spheres, concluding that our planet would have cooled to present temperatures in 42,964 years and 221 days.

After a number of attempts using different parameters, Lord Kelvin obtained an age of about 20 million years, which did not convince the geologists and made Darwin unhappy. Darwin's evolutionary theory rested on a much larger figure. The geologists attempted to upstage the physicists, using geological processes to calculate the age of the Earth. They used the dissolution of salt in the oceans to get an age of the planet. They assumed that all the salt present in the oceans today derived from rock decomposition and was delivered by the rivers at a certain rate. They obtained an age of about 90 million years for the formation of the first ocean waters. The geologists then added a couple of hundred million years using the accumulation of the deposits of the rock strata, assuming a known rate of sediment deposition. But there were too many assumptions and nobody was convinced of these results.

The physicists returned to centre stage at the turn of the 19th century with their revolutionary discoveries on radioactivity and atoms. The geologists, who were convinced that the Earth was much older than commonly thought, where glad to hear that radioactivity was heating materials inside the Earth while the planet was cooling down. This could have explained the relatively young age calculated by Kelvin and supported the ideas of an older Earth. In reality, Kelvin's young Earth was not resulting from the contribution of radioactivity to its heat budget: his main error was the assumption of a solid Earth. In 1885, John Perry, an Irish scientist who had been Kelvin's assistant, argued in a *Nature* paper that the Earth's interior was fluid, with the heat being distributed by convection. A small solid outer skin would keep the temperature gradient at the surface high for a long period. On the basis of these assumptions, Perry evaluated an age of 2 to 3 billion years for the Earth.

Rutherford himself contributed to the above misunderstanding on the role of radioactivity in Kelvin's calculations. On the other hand, the New Zealander was the first to use uranium radioactivity to measure the age of rocks. He used the accumulation of helium that he and Soddy had identified as one of the radioactive decay products of uranium. Helium was always found in uranium-rich minerals. In 1905, measuring its concentration in a sample of pitchblende, he evaluated that the mineral was 500 million years old.

Unfortunately, the new dating method had its limits. Helium was a gas and could be partially lost during the analysis, escaping from the crushed rock. Rutherford's helium-based methods only provided an estimate of the minimum age of minerals.

Meanwhile, the American scientist Bertram Boltwood had determined that lead was also ever-present in uranium minerals. He correctly assumed that lead was the final product of uranium.

The first to apply the uranium-lead dating method was the geophysicist Arthur Holmes, who obtained, in 1911, an age of 370 million years for a rock from Norway attributed to the Devonian. He continued his work dating other rocks of different geological periods, establishing the first geological time scale, back to 1.6 billion years ago.

The main problem in the application of the uranium-lead method was related to the possible presence of lead that was not deriving from the decay of uranium. For example, the lead in the rock could be the product of thorium decay; or it could already be present in the mineral from the time of its formation. In this last case, it is called *ordinary* lead.

In 1927, Aston determined that lead has three isotopes: lead-206 (24.1%), lead-207 (22.1%), and lead-208 (52.4%). He and Rutherford developed the idea that the isotope of lead considered of primordial origin, lead-207, was instead the product of a second isotope of uranium, uranium-235, previously unknown. In 1936, the American physicist Alfred Nier, using a more advanced spectrometer, found that the primordial lead was another isotope, lead-204 (only 1.4%). Eventually, the identification of all the different lead isotopes, including those deriving from uranium decay, led to the development of a precise uranium-lead dating method.

Aston also deduced the ratio between uranium-235 and uranium-238, allowing Rutherford to calculate an age of 3,400 million years for the Earth, under the assumption that the two isotopes had the same abundance when the planet formed.

Now the uranium-lead method could help advance the dating of the Earth, as there were two decays that could be exploited: that of uranium-238 to lead-206, and that of uranium-235 to lead-207. The age of Earth could also be determined with high precision from the ratio of the uranium products, lead-206 and lead-207, to lead 204, which is fixed. The use of this method allowed Nier to obtain ages of more than 2 billion years for some igneous rocks – older than the age of 1.8 billion years assigned then by the astronomers to the universe, using the first measurements of the recession velocities for Andromeda and other galaxies. It will be shown in the 1950s that Hubble had assumed the wrong distance for the galaxies. His contribution to cosmology was recognized in any case by the Nobel physics committee, but he died on 28 September 1953, while his nomination was being prepared.

The final problem to solve for evaluating the age of the Earth was to measure the composition of primordial lead to determine the contribution of radiogenic lead from the decay of uranium. The solution to this problem came from the meteorites.

Iron meteorites have a small amount of uranium, so their lead should be prevalently of primordial origin. In 1953, the American geochemist Clair Patterson measured the isotopic composition of lead in the Canyon Diablo meteorite, the body that created the Meteor Crater in Arizona 50,000 years ago. In 1956, he published his famous 'isochron' graph, showing that the lead-207/lead-204 versus lead-206/lead-204 ratios of meteorites and ocean floor samples were on the same straight line. The gradient of the line

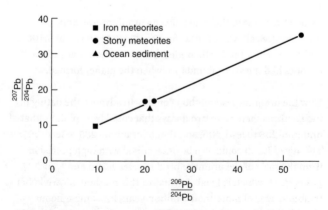

25. The lead isochron for ocean sediment and meteorites. The slope indicates an age of 4.55 ± 0.07 billion years

gave the age of these materials – 4.55 billion years with an error ± 70 million years. The race to calculate the age of the Earth was over.

Remains of life

Materials belonging to organisms that were alive during the deep past have been known since ancient times, often becoming part of myths and religious beliefs. They were thought to be, for example, the bones of animals that perished during the biblical deluge, or the remains of dragons or those of fantastic demons. Of course, there were also more realistic interpretations. Already 500 years ago, Leonardo da Vinci interpreted fossils as the products of organisms that once lived in the sea. With the advent of the so-called period of Enlightenment, reasoning prevailed and a growing number of naturalists interpreted fossil remains as the leftovers of creatures that had populated the planet in its deep past. Darwin himself used the fossils of extinct animals, such as the giant sloths of Bahía Blanca in Argentina, to prove that species are not immutable. Palaeontologists have been studying ancient fossilized bones for centuries, attempting to understand what occurred to the original organic material, as it was transformed into a mineral. It was accepted that the original living tissues had been fully transformed into minerals by fossilization. But modern science is giving the option of looking at these old remains with new eyes.

With their sophisticated tools, physicists have recently shown that there is a chance that fossil remains, many millions of years old, could preserve part of the original material. Novel synchrotron radiation X-ray fluorescence microscopes provide images of trace and minor elements in the sample.

The method was applied to a fossil *Archaeopteryx*, a 145-million-year-old proto-avian theropod dinosaur. It revealed phosphorus, zinc, and copper, key elements in modern birds, proving that chemical elements that were in the original tissues of the animals can still be found in the fossil remains.

Scientists noticed that the feathers of fossilized dinosaurs had structures called melanosomes. They contain melanin that gives the feathers of living birds their colour. Scientists want to use these advanced techniques to find out the colour of tail feathers of the flying dinosaur *Sinosauropteryx* that lived in China 125 million years ago.

Another analytical technique, synchrotron radiation computed micro-tomography, can be used to peer inside dinosaur eggs and look for embryos. The images of the tiny dinosaur bones can be 'virtually' extracted from the egg with resolution down to a thousandth of a millimetre or less. The method also opens the way to virtually section bones of fossilized skulls, to trace the microstructure of the original tissue, and find out if some dinosaur species were head-butters or had head crests. The technique could also reveal lines of arrested growth, common in the bones of modern amphibians and mammals, corresponding to periods of slow biologic development. The dinosaur lifespan could be evaluated by counting these lines in non-invasive ways using micro-CT imaging.

Dinosaurs living at the end of the Cretaceous were wiped out by the KT disaster, 65.5 million years ago.

After the dinosaurs

In the beautiful Bottacione valley near Gubbio, the KT boundary – the thin layer of clay that shows the transition between the Cretaceous and the Tertiary periods – is part of the touristic attractions. This geologic site became famous in 1980 after the American scientist Luis Alvarez, the 1960 Nobel laureate in physics, and his geologist son, Walter, analysed the KT sediments with neutron activation analysis. They identified an anomalous spike of iridium in the KT clay layer, 3,000 parts per trillion against the 10 to 15 parts per trillion found in normal terrestrial sediments. Since meteorites and other extraterrestrial materials have high concentrations of iridium, this could be interpreted as the effect of a 10-kilometre asteroid hitting the Earth some 65.5 million years ago. Together with a dramatic drop in plant and animal biodiversity at planetary scale, the dinosaurs were wiped out by this environmental disaster. Now it is widely believed that the impact formed the 180-kilometre-diameter Chicxulub crater in the Yucatan Peninsula, Mexico. Discovered in the 1970s, this crater has in fact the same age as the iridium layer of the KT boundary, which has been identified in many parts of the world. The identification of tektites, shocked quartz, and gravity anomalies demonstrate its impact origin.

Above the white chalk of the Cretaceous and the KT clay, lie the strata that correspond to the last layers of the planetary geological skin. The Tertiary starts 65.5 million years ago and leads to the Quaternary, our period, beginning a mere 2.6 million years ago.

About 50 million years ago, a global cooling trend took our planet from the tropical conditions at the beginning of the Tertiary to the ice ages of the Quaternary, when the Arctic ice cap developed. The temperature decrease was accompanied by a decrease in atmospheric CO_2 from 2,000 to 300 parts per million. The cooling was probably caused by a reduced greenhouse effect and also by changes in ocean circulation due to plate tectonics. The drop in temperature was not constant as there were some brief periods of sudden warming. Ocean deep-water temperatures dropped from 12 °C, 50 million years ago, to 6 °C, 30 million years ago, according to archives in deep-sea sediments (today, deep-sea waters are about 2 °C).

Global climate change was the engine of the selective process that led to the differentiation and spread of the primates.

During the Tertiary, the Gondwana supercontinent split, driving India and the Eurasian plate onto a collision that created the Himalayas and the Tibetan Plateau Antarctica attained its present position at the South Pole, and South and North America reconnected, following the rapid expansion of the Drake channel. The Himalayan plateau changed atmospheric circulation and its freshly exposed rocks absorbed carbon dioxide, speeding up the cooling of the planet and harshening the landscapes in many regions.

Recent discoveries suggest that the first apes appeared nearly 30 million years ago in the area of present-day Saudi Arabia – at the time still connected to Africa. At about the same geologic time, the Arabian plate broke off from the African Shield and started sliding to the north-north-east, towards Eurasia, rotating counter-clockwise. About 19 million years ago, Eurasia and Africa were eventually connected. The planet warmed up again for a period of a few million years, with an increase in vegetation and forests; as a result, the apes could spread to the new lands.

For millions of years, a large variety of apes (more than a hundred species of hominoids) occupied an area extending from the Iberian Peninsula to eastern Asia and southern Africa, with the sea periodically separating and reconnecting Eurasia and Africa. The cooling trend restarted in the final part of the Tertiary, during the early Pliocene.

In spite of being called the 'Ice Age', the following epoch (the Pleistocene), which spanned from 2.6 million to 12,000 years ago, was characterized by intermittent glacial advance and recess. During the last 2 million years, the mean duration of the glacial periods was about 26,000 years, while that of the warm periods – interglacials – was about 27,000 years. Between 2.6 and 1.1 million years ago, a full cycle of glacial advance and retreat lasted about 41,000 years. During the past 1.2 million years, this cycle has lasted 100,000 years.

Stable and radioactive isotopes play a crucial role in the reconstruction of the climatic history of our planet that was briefly outlined above. During the most recent geological periods, global and regional environmental changes set the stage for the appearance of numerous species of apes, including our own variety. The history of the past environment is written in ice cores, marine sediments, spelothems, corals, and tree rings. The history of the apes – both extinct and in existence – is written in their fossil bones and in their DNA; the naked ape also left traces of its 'culture'. It is precisely to these arguments that we turn, in the next and final chapter, to conclude our journey through radioactivity and its applications.

Chapter 8
Tracing human origins and history

A long time ago, in a place now called Lateoli, in northern Tanzania, a group of bipedal creatures, two adults and a child, walked over wet volcanic ash that later hardened like concrete. Their footprints were found in 1978 by the British archaeologist Mary Leakey.

The three hominins, perhaps a family-like group, were later identified as members of the genus *Australopithecus*, the branch of the human tree that immediately precedes our own genus, *Homo*. They weighed 30 to 50 kilograms and stood a little over one metre tall. The australopithecines had a cranial capacity of 400 to 500 cubic centimetres, close to that of a chimpanzee.

The first *Australopithecus* fossil, the Taung Child, was discovered in South Africa in 1925 by Raymond Dart, of the University of Witwatersrand. This species was the first evidence corroborating Darwin's original idea that the ancestors of the human family probably originated in Africa. Several other australopithecine species have been discovered since Dart's find, including the famous *A. afarensis*, Lucy, from Ethiopia.

When were these early hominins living? Who were their possible ancestors? How did they evolve? Are they really our ancestors? Many answers to these questions come from using radioactivity-based geochronometers and advanced radiation-based microscopes.

The first hominins

Volcanoes have erupted frequently in the Rift Valley of East Africa. In fact, plate tectonics had torn Africa apart for more than 40 million years. The volcanic ash, interlayered with fossils, contains materials that can be dated using radioactivity-based methods. Volcanic glasses in East Africa have a high concentration of potassium; the rate of decay of the unstable isotope potassium-40 to argon-40 can be used to date the ash. The radiogenic argon is all released to the atmosphere when the volcano erupts, so that the clock is reset as the volcanic ejecta begin to cool down and argon-40 starts accumulating.

The concentration of argon-40 can be measured by heating the mineral sample, using a mass spectrometer to count the atoms thus released. Unfortunately, we need to know the relative concentrations of both the parent and the daughter isotopes to work out the sample's age. An elegant method was developed in the 1960s at the University of California, Berkeley, to analyse potassium concentration. The sample is irradiated in a reactor, where nuclear reactions of neutrons with potassium produce argon-39, which can be used as a proxy of potassium concentration. Argon-39 and argon-40 can then be simultaneously analysed with mass spectrometry. Hence this method is called 'argon–argon dating'.

Dating accuracy has been enhanced by extracting the argon atoms – using a laser – from single crystals picked from the pumice. The method works best on samples that are at least 100,000 years old, but it can be used on samples as young as 10,000 years as long as they are rich in potassium. It can also be extended to chronologies close to the age of the Earth.

Potassium-40 radioactivity gives reliable dates for the key turning points in human evolution in East Africa. The australopithecine family left its footprints in the ash 3.6 million years ago, while Leakey's *Australopithecus boisei* and the *Australopithecus afarensis*, Lucy, are found to be 1.75 million and 3.18 million years old, respectively.

Another much older bipedal female was discovered in the Afar depression, in Ethiopia. Some consider her to be the predecessor of Lucy. The bones of the *Ardipithecus ramidus*, Ardi for the media, were bracketed by strata deposited by two volcanic eruptions; they allow us to give her an argon–argon age of 4.4 million years. High resolution X-ray CT scanning of Ardi's bones suggests a 'multicomponent' locomotory behaviour. Even if she evolved shortly after the common ancestor of chimpanzees and humans, she neither knuckle-walked nor swung in the tree canopy. Her limbs had a combination of features, which allowed her both to climb the trees and, at the same time, explore new habitats walking upright on the ground.

The teeth of an earlier hominin, *Ardipithecus kadabba*, found in the same part of Ethiopia, have been dated through potassium radioactivity to 5.6 million years. *A. kadabba* is very similar to other hominins such as *Orrorin tugeniensis* from Kenya, with a potassium–argon age of about 6 million years, and *Sahelanthropus tchadensis*, from Central Africa, of 6 to 7 million years. This age, which was originally obtained using 'biochronological' methods (i.e. based on mammalian fossils of known chronology, co-located with the hominin remains), was recently confirmed using cosmogenic dating with the beryllium-10/beryllium-9 method.

Beryllium-10 dating is similar to the radiocarbon dating that we discussed in Chapter 5. Beryllium-10 is produced in the atmosphere by high-energy cosmic rays bombarding oxygen and nitrogen. The radionuclide is then adsorbed in aerosols and transferred to the Earth's surface in soluble form by precipitation. Finally, beryllium-10 is associated with sediments, where it decays. The ratio between beryllium-10 and beryllium-9 – assuming that the latter, present in the sediments at trace levels, has been homogenized to beryllium-10 – works as a geo-chronometer. The two sediment layers bracketing the *S. tchadensis* cranium provide beryllium-10/beryllium-9 dates of 6.83 and 7.12 million years, respectively, confirming the earlier age estimates.

These early hominins have ages consistent with what emerges from genetic studies, showing that the common ancestor of chimpanzees and modern humans lived around 5 to 7 million years ago. The distorted skull of *S. tchadensis* was virtually reconstructed using imaging techniques based on high-resolution computed X-ray tomography. The analysis of his basicranium suggests that this species might have been an upright biped.

South African australopithecines cannot be dated with argon–argon since there is no volcanic material available. 'Little Foot', the *Australopithecus* found by the palaeoanthropologist Ron Clarke in 1994, in the Sterkfontain cave in South Africa, was dated using cosmogenic 'burial dating' based on the beryllium-10/ aluminium-26 method.

Beryllium-10 and aluminium-26 are produced by secondary cosmic rays (mainly neutrons and muons, in turn generated by the irradiation of the upper atmosphere by high-energy protons) bombarding silicon minerals at the Earth's surface. When the minerals are buried at depths of several metres, out of the reach of neutrons and partially shielded from muons, radioisotope production nearly stops. Radioactive decay decreases the concentration of the radionuclides, so they work as chronometers, measuring the time elapsed since the material was buried. The residual concentration of beryllium-10 and aluminium-26, measured with AMS atom counting, provides 'burial ages' for the rocks in the cave back to some million years ago.

The resulting burial age for Little Foot was 4 million years, but the number was at odds with palaeomagnetic dating.

In 2006, a near-complete young *Australopithecus afarensis* was found at Dikika, Ethiopia. Electron and X-ray imaging of animal bones found nearby show the marks of stone tools. The age of the bones, based on argon–argon ages of the bracketing tuff, is 3.39 million years. Once confirmed, this should be the earliest evidence for meat and marrow consumption in the hominin lineage. Until recently, it was thought that this behaviour was developed by later hominins, around 2.5 million years ago, as shown by the earliest stone tools from the Gona region in Ethiopia.

Australopithecus was a very successful genus, and survived between 4 and nearly 1 million years ago, with the most robust taxa. It weathered extreme environmental change in Africa, as geological and astronomical forces deeply transformed Earth's climate at the time.

At the onset of the Pleistocene, Africa was becoming increasingly dry and cold, with devastating effects on its landscape. Forests gave way to savannas and open terrain. Many forest-adapted animal species did not survive. Evolution's selective forces shaped new animal species, including those in the Hominini tribe. *Australopithecus boisei* was one of them, appearing in the fossil record less than 2 million years ago. Soon nicknamed 'Nutcracker Man', he had flatter cheek teeth than those of earlier australopithecines, with thick enamel and robust mandibles. This is consistent with a change of diet: from soft fruits and leaves to nuts, seeds, roots, and tubers. A similar hominin, *Australopithecus robustus*, appeared in South Africa. It had powerful chewing muscles, anchored to the characteristic sagittal crest, and large molar teeth with thick enamel, suitable for eating tough, fibrous plants. Several paleoanthropologists, including Colin Groves from the Australian National University, do not accept that 'robust australopithecines' (*boisei* and *robustus*) should actually be placed in the genus *Australopithecus*. Most would put them in the genus *Paranthropus*. But Colin is said to be a 'lumper': he would put all species of hominins as far back as *Ardipithecus* into the genus *Homo*, as also suggested by some geneticists.

Homo

The forces of climate and environmental change during the Pleistocene speeded up the engine of natural selection, modifying the diet and the general way of life of the tribe Hominini. The new prevailing conditions demanded increasing degrees of adaptability. The fossil record shows that a second direction was taken during this critical period, with the appearance of a hominin characterized by a slightly larger brain, up to about 600 cc.

Remains classified as *Homo habilis* show ages of between 2.3 and 1.4 million years. This *Homo*'s arms were well developed compared to his legs, suggesting that he still maintained considerable climbing abilities. His body size was comparable to that of *Australopithecus*.

Stone tools assigned to *H. habilis*, found in Hadar, Ethiopia, were dated with argon–argon and fission-track dating techniques. The latter is based on the accumulation of damage in volcanic rock or glass from the fission of uranium impurities. When uranium-238 nuclei spontaneously fission, the resulting fragments leave a micrometre-sized track in the mineral, which is preserved for millions of years, as long as the temperature of the mineral stays below a certain level reached only in violent geologic processes. The density of tracks, made visible by chemical etching, indicates the time elapsed since the last volcanic eruption. Age evaluation requires knowing uranium concentration. This can be obtained by irradiating the specimen in a nuclear reactor. Neutrons induce fission of uranium-235, producing further tracks that can be used to estimate uranium concentration in the mineral. The results of these analyses suggest an age of 2.5 million years for the Hadar stone tools.

South African palaeoanthropologists recently discovered, in the Malapa cave, near Johannesburg, a new hominin, *Australopithecus sediba*, dated by the uranium-lead method to 1.977 ± 0.002 million years ago. It has a small brain and other australopithecine characters, but its teeth, legs, and pelvis evoke *Homo*. Synchrotron radiation microtomography of the cranium allowed reconstructing the shape of its brain, characterized by slightly asymmetric frontal lobes, similarly to humans. Results suggest that neuron reorganization started in pre-human lineages before the expansion of their brain, with a possible increase of connections in the region that in humans is associated with language and social behaviour. The discoverers believe that this species was transitional between *A. africanus* and *Homo*. Others are not convinced that *Homo* necessarily evolved from any of the *Australopithecus* species we know.

A new species, *Homo ergaster*, appeared suddenly in Africa about 2 million years ago. Nariokotome Boy was 1.75 metres tall, had almost twice the weight of an *Australopithecus*, and a brain capacity of 900 cc. Synchrotron radiation microtomography was used to determine his age at death, studying the microstructure of one of his teeth. In spite of his body size, the Nariokotome Boy was only 8 years old, showing that this hominin, like chimps, matured early. His teeth and the bones of the skeleton suggested different ages at death, swinging between 8 and 14 years. This implies that the lengthening of the growth period was not a simple transition. *H. ergaster*, with its smaller jaw muscle and smaller molar teeth, had a diet of softer food, including fruits and meat.

The very earliest Acheulian tools (found at Kokiselei, West Turkana) are dated at 1.76 million years, within the time span for *H. ergaster*. Acheulian tools were made using pebbles, from which a few flakes were struck off using other stones. Most scholars believe that the technology used by *H. ergaster* was the so-called 'Developed Oldowan' (pebble tool, or Mode 1). The earliest human traces outside Africa have Mode 1 technology, while hand axes (Acheulian, Mode 2) appear later.

According to some investigators, *H. ergaster* had tamed fire. This might have been the disruptive weapon that allowed *ergaster* to 'terrorize' the last surviving species of *Australopithecus*, *boisei* and *robustus*, and its fellow human species such as *habilis* and *rudolfensis* in the savannahs of Africa.

Out of Africa again and again

H. ergaster was the first species, as far as we know, to disperse outside Africa, advancing through the vast savannahs connecting, at the dawn of the Pleistocene, the Rift Valley to eastern Asia. *H. erectus*, its Asian variant, was characterized by a thick skull, flat forehead, and projecting face, well represented by the crania from the site of Sangiran on Java, dated to about 1 million years ago. Another example of Indonesian *H. erectus* is the Modjokerto child cranium, also from Java, which has an argon–argon date of 1.8 million years.

The Dutch medical doctor Eugene Dubois discovered the first Indonesian *erectus* fossils in 1891 at Trinil, on the banks of the Solo River. He was convinced they were the remains of a creature somewhere between humans and apes, the so-called 'missing link', a misleading concept that became popular after the publication of Darwin's theories on evolution, and was perpetuated by the media.

In fact, Dubois called the Trinil hominin *Pithecanthropus erectus*, 'ape-human that stands upright', also known as Java Man. The discovery supported a possible Asian human cradle, contradicting the ideas of an African origin proposed by Darwin 20 years before in his book *The Descent of Man*.

In the 1930s, other *erectus* remains were found at Zhoukoudian near Beijing. They were attributed to *Sinanthropus pekinensis*, or

Peking Man. In 2009, beryllium-10/aluminium-26 burial dating provided ages in the range 780,000 to 680,000 years for Peking Man, 200,000 years older than previously thought. In the 1950s, both Java Man and Peking Man were lumped together into the species *H. erectus*.

Remains of hominins considered *habilis/ergaster* or *erectus*-like and also referred to as *H. georgicus* were found in Dmanisi, Georgia, in the 1990s. They have potassium–argon dates of 1.8 million years. So far, the relationship among *H. georgicus*, the Asian *erectus*, and the African *ergaster* is not fully understood. Some of the Dmanisi remains can be linked to the primitive end of the *H. erectus/ergaster* range, with one of the crania having a brain size of only 600 cc. It has been suggested that the ancestors of Dmanisi hominins were the earliest migrants from Africa.

About one million years ago, archaic human groups inhabited a vast area, from the island of Flores in Indonesia to the Iberian Peninsula. Argon–argon dating was applied to the volcanic material overlaying artefacts found in central Flores, giving one million years as the minimum age for the human presence on the island. In Europe, *H. antecessor* is one of the oldest human taxa. A mandible, found at the Sima del Elefante cave site in Spain, was dated with the beryllium-10/aluminium-26 method to 1.2 to 1.1 million years ago, confirming palaeomagnetic and biostratigraphic dating.

The Middle Pleistocene in Europe preserves many fossil remains of the archaic *Homo* lineage. They have been attributed to a variety of different species, such as *cepranensis*, *heidelbergensis*, *petralonensis*, and *steinheimensis*. These attributions are not endorsed by most palaeoanthropologists. Furthermore, the remains are found in sites with uncertain chronology.

The interpretation of the human fossil record is based on two main evolutionary models. The multiregional model claims that modern humans evolved independently in different continents from more archaic species. Gene flow between populations pushed a single species down the same evolutionary pathway. According to the rival 'out of Africa' model, modern humans all have a recent African ancestry. *H. sapiens* evolved in Africa about 200,000 years ago and migrated to other continents, replacing more archaic hominins, descendants of earlier migrants.

Recent multiregional theories endorse an assimilation mechanism. According to this approach, archaic Chinese humans evolved from Peking Man and assimilated some African-related migrants. The resulting *Mongoloids* dispersed later into the new world. Archaic Java *H. sapiens* interbred with the Africans to produce the *Australoids*, who dispersed into Australia. The Africans interbred with the Neanderthals, originating the *Caucasoids* in Europe and West Asia. To the multiregionalists, the different European species of the Middle Pleistocene were all variations on the *sapiens* species, while to the supporters of the 'out of Africa' model, the original European stock represented evolutionary dead ends.

The extreme form of the 'out of Africa' model was given a blow in 2010, with the publication of a draft sequence of the Neanderthal genome.

Many palaeoanthropologists think that the Neanderthals evolved perhaps as early as 400,000 years ago from *H. heidelbergensis*, a descendant of African *H. ergaster* who migrated to Europe. The *H. heidelbergensis* populations that remained in Africa (called *H. rodhesiensis*) evolved into *H. sapiens*.

Surprisingly, the Neanderthals' nuclear DNA recently showed that these hominins contributed up to 4% of the DNA of modern humans outside Africa. These results are based on the analysis of 60% of the *H. neanderthalensis* genome performed on 43,000–47,000-year-old bones from Vindija cave in Croatia. The

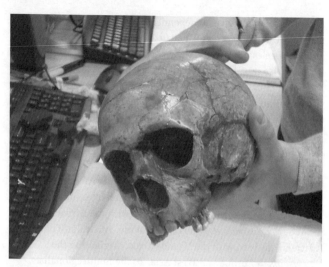

26. Skull of a Neanderthal child. The fossil found at the site of Pech de l'Aze belongs to a Neanderthal child who was two to three years old at death

results challenge the extreme version of the out of Africa model, which states that there was no 'gene flow' between Neanderthals and the African *H. sapiens* migrants.

Although there is a broad consensus on the genesis of the modern human anatomy, which dates back to about 200,000 years ago, according to both fossils and population genetics, when the 'modern mind' developed is much more controversial. Indeed, the question when modern humans turned from 'anatomically' modern into 'behaviourally' modern still puzzles scholars.

Again, radioactivity methods can determine the chronology of events associated with different human behaviour.

Until recently, it was widely accepted that modern human behaviour emerged during the so-called 'Upper Palaeolithic

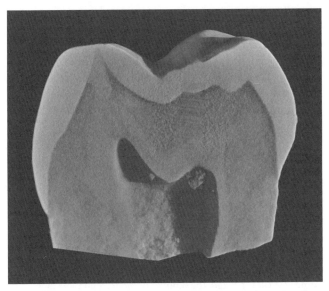

27. X-ray microtomography of a Neanderthal tooth performed at the Tomolab, Sincrotrone Trieste, Italy. The virtual section shows the detailed structure of enamel, dentine, and pulp chamber

revolution' in Europe, 40,000 to 35,000 years ago. This hypothesis was shaken some years ago when perforated shells and pieces of ochre with graphic symbolism were found in the Blombos cave, South Africa, showing that human culture had probably developed much earlier: at least 70,000 years ago.

In addition, advanced stone artefacts – called Still Bay points – were found in the same region, showing innovative techniques that were not present in previous periods. A second period of innovation is represented by another lithic industry, called Howieson's Poort. Recent optically stimulated luminescence dating shows that the Still Bay technology lasted only 1,000 years, around 71,000 years ago, while the Howieson's Poort industry lasted 5,000 years, starting around 65,000 years ago. OSL dating is based on the energy accumulated by crystals, such as silicon and feldspar in sediments, when they absorb ionizing radiation emitted by uranium, thorium, potassium, and cosmic rays. The energy stored is proportional to the time elapsed since the last exposure of the crystals to sunlight, which erases the luminescence signal, resetting the 'clock'. Zapped with light, the stored energy is evicted, causing luminescence, the intensity of which is a measure of age.

Archaeological studies in southern Africa and population genetics show that a period of technological innovation corresponded to population increase. Some scholars believe that it was probably during one of these population increases that modern humans expanded out of Africa, with their baggage of lithic innovation, complex language, and symbolic thought. Others develop different models, related to population bottlenecks.

Some 70,000 years ago, the number of modern humans had diminished to a few thousand individuals, as a probable result of the environmental disaster caused by the Toba eruption. The eruption of the Toba volcano in Sumatra, 74,000 years ago, was probably the biggest volcanic explosion in the past 2 million years; it threw 2,700 cubic kilometres of ash into the atmosphere, causing global climate change. The disaster might have contracted the tropical forest to isolated *refugia*. *H. sapiens* was possibly forced to explore sea travel.

Many scholars believe the exodus of *H. sapiens* from Africa, more than 1.7 million years after the first dispersal of *Homo ergaster/ erectus* from the same continent, contributed to the extinction of the four known contemporary human taxa, probably present in small numbers: the last *erectus*-like groups and *H. floresiensis* in Asia, *H. neanderthalensis* and the Denisovans in Eurasia. The latter is an enigmatic hominin whose remains, the distal manual phalanx of the fifth finger and a tooth, were found in the Denisova cave in southern Siberia. The DNA analysis of the phalanx, found in a sediment layer dated between 48,000 and 30,000 years ago, shows that this individual, a female, was different from both Neanderthals and modern humans. The genome also shows that the Melanesians inherited at least one-twentieth of their DNA from the Denisovans.

DNA analysis in present-day human populations suggests that modern humans left Africa around 70,000 years ago, reaching Australia, generation after generation, around 50,000 years ago, Europe 40,000 years ago, and America 13,000 years ago.

The archaeological record confirms this scenario. During the oxygen isotope stage 3 (OIS3), the signs of complex social behaviour and broad use of symbolic expressions appear first in Africa, then in Australia, Eurasia, and elsewhere. Outstanding variations of that ancient culture are witnessed by the funerary burial remains of the cremated Mungo Lady and the buried Mungo Man in Australia, dating over 40,000 years ago, and by the bone musical instruments and the rock art in Europe, 35,000 years ago.

Much work is needed, though, to confirm the details of human dispersal. The use of precise radioactivity-based chronometers, such as the ones we discussed above, including radiocarbon, OSL, U-series, and ESR dating, will keep providing useful information in the years to come.

Studies on human origins need reliable dates to answer thorny questions about the dispersal of modern humans during the late Pleistocene, including their impact on the ecosystems. Their possible involvement in the extinction of the megafauna in Australia and America is particularly controversial, fuelling arguments over the 'politics of the past'.

Several Australian groups are re-evaluating archaeological sites to find out the period of overlap of people and megafauna, in a bid to discriminate between human and climatic causes of extinction of the big animals such as the wombat-like diprotodon and the giant kangaroos. Present results point to a human/megafauna overlap of a few thousand years in Australia. In the Americas, the period of overlap between the Clovis hunters and megafauna, such as the sabre-toothed tiger, was also relatively short.

As we know, the impacts of modern humans in the Holocene – the current warm period that started at the end of the last Ice Age – has not been less impressive. A recent discovery offers a unique insight on the behaviour of modern humans, just in the middle of this period.

One species, many individuals

On 19 September 1991, a frozen body was found trapped in a glacier in the Ötzaler Alps of the Austrian–Italian border. It was well preserved, and initially thought to be the remains of a mountaineer who died some 10 or 20 years before. But the radiocarbon analyses on his body, clothing, and implements revealed that this man, soon called Ötzi, had died between 5,300 and 5,100 years ago.

A flurry of forensic studies were developed to pinpoint Ötzi's origins and to understand what happened in the last days of his life. X-ray computed tomography scans revealed what looks like an arrowhead planted in Ötzi's left shoulder, implying that he may have been involved in a violent fight, perhaps the cause of his death.

He was one of the few million modern humans that inhabited the world at the end of the Neolithic. Before the 'Neolithic Revolution' took place, some 10,000 years ago, with the development of agriculture and the domestication of animals, the number of humans did not exceed one million individuals. Recent DNA studies on skeletons found in Germany, radiocarbon dated to about 7,000 years ago, revealed that the agricultural revolution that allowed for this strong population increase took place, in Europe, only after the settling of new groups from the Middle East, whose population had boomed precisely for the same reasons: the initial surpluses delivered by agriculture. These results connect the lineage of the first European farmers to the populations in present-day Iraq, Syria, and nearby countries.

No other species can now compete with modern humans, who consider themselves at the top of all other creatures and keep growing in number and in the use of increasingly scarce resources. The discovery of radioactivity and its applications improved our survival kit, but also gave us the chance to reach a new level of awareness on the history of our species. The new challenge, from now on, will be to learn the most from our deep past, to better face our uncertain future.

References

Chapter 1: Opening the nuclear Pandora's box

J. P. Adloff, 'The Laboratory Notebooks of Pierre and Marie Curie and the Discovery of Polonium and Radium', *Czechoslovak Journal of Physics*, 49 (1999): 15–28.

Malcolm C. Henderson, M. Stanley Livingston, and Ernest O. Lawrence, 'Artificial Radioactivity Produced by Deuton Bombardment', *Physical Review*, 45 (1934): 428–9.

M. Abid, A. Begun, A. S. Mollah, and M. A. Zaman, 'Measurements of Radioactivity in Books and Calculations of Resultant Eye Doses to Readers', *Health Physics*, 88 (2005): 169–74.

<http://www.world-nuclear.org/info/inf05.html> (accessed March 2012)

Chapter 2: Unlimited energy

H. A. Bethe, 'Energy Production in Stars', *Physical Review*, 55 (1939): 434–56.

IAEA, 'Thorium Fuel Cycle – Potential Benefits and Challenges', IAEA-TECDOC-1450 (2005).

Chapter 4: Radiation and radioactivity in medicine

J. Walton, C. Tuniz, D. Fink, G. Jacobsen, and D. Wilcox, 'Uptake of Trace Amounts of Aluminium into the Brain from Drinking Water', *Neurotoxicology*, 16 (1995): 187–90.

Chapter 5: Radioactive gadgets and gauges

<http://www.epa.gov/radiation/source-reduction-management/applications.html> (accessed March 2012)

Chapter 6: Fear of radioactivity

M. A. C. Hotchkis, D. Child, J. Ferris, and C. Tuniz, 'Application of Accelerator Mass Spectrometry for Uranium-236 Analysis', *Journal of Nuclear Science and Technology*, 3 (2002): 532.

IAEA, 'Nuclear Security Measures at the XV Pan American Games: Rio De Janeiro 2007', IAEA, Vienna (2009).

U. Zoppi, Z. Skopec, J. Skopec, C. Tuniz, et al., 'Forensic Applications of C-14 Bomb-Pulse Dating', *Nucl. Instr. and Methods in Phys. Res. B*, 223 (2004): 770–5.

S. Thomson, M. Reinhard, M. Colella, and C. Tuniz, 'Unmasking the Illicit Trafficking of Nuclear and Other Radioactive Materials', in *Radionuclide Concentrations in Foods and Environment*, ed. N. Nollet and M. Poschl (New York: Marce Dekker, 2006), pp. 333–65.

Chapter 7: Tracing the origin and the evolution of Earth

<http://www.physicsoftheuniverse.com/topics_bigbang_timeline.html>

<http://en.wikipedia.org/wiki/Timeline_of_the_Big_Bang#Hadron_epoch>

<http://math.ucr.edu/home/baez/timeline.html#bang> (all accessed March 2012)

Abderrazak El Albani et al., 'Large Colonial Organisms with Coordinated Growth in Oxygenated Environments 2.1 Gyr Ago', *Nature*, 466 (2010): 100–3.

F. J. Vine and D. H. Matthews, 'Magnetic Anomalies over Ocean Ridges', *Nature*, 199 (1963): 947–9.

C. Tuniz, D. K. Pal, R. K. Moniot, W. Savin, T. H. Kruse, and G. F. Herzog, 'Recent Cosmic Ray Exposure History of ALHA81005', *Geophysical Research Letters*, 10 (1983): 804.

S. Török, K. Jones, and C. Tuniz, 'Characterisation of Minerals Using Ion and Photon Beam Methods', in *Nuclear Methods in Geology*, ed. A. Vértes et al. (London: Plenum Press, 1998), pp. 217–49.

D. K. Pal, R. K. Moniot, T. H. Kruse, C. Tuniz, and G. F. Herzog, 'Spallogenic Be-10 in the Jilin Chondrite', *Earth and Planetary Science Letters*, 72 (1984): 273.

J. Chela-Flores, M. E. Montenegro, N. Pugliese, V. Tewari, and C. Tuniz, 'Evolution of Plant–Animal Interaction', in *All Flesh is Grass: Plant–Animal Interactions, A Love–Hate Affair*, ed. J. Seckbach and Z. Dubinsky; reprinted in *Cellular Origin and Life in Extreme Habitats and Astrobiology* (Dordrecht: Springer, 2009).

J. Chela-Flores, G. Jerse, M. Messerotti, and C. Tuniz, 'Astronomical and Astrobiological Imprints on the Fossil Records: A Review, in *'From Fossils to Astrobiology'*, ed. J. Seckbach, in *Cellular Origins, Life in Extreme Habitats and Astrobiology* (Dordrecht: Springer, 2009), pp. 389–408.

P. Carlson and A. de Angelis, 'Nationalism and Internationalism in Science: The Case of the Discovery of Cosmic Rays', *Eur. Phys. J. H*, DOI: 10.1140/epjh/e2011-10033-6

P. C. England, P. Molnar, and F. M. Richter, 'Kelvin, Perry and the Earth', *American Scientist*, 95 (2007): 342–9.

Chapter 8: Tracing human origins and history

R. Pickering et al., *'Australopithecus sediba* at 1.977 Ma and Implications for the Origins of the Genus *Homo'*, *Science*, 333 (2011): 1421.

K. J. Carlson et al., 'The Endocast of MH1, *Australopithecus sediba'*, *Science*, 333 (2011): 1402.

Matt Sponheimer et al., 'Isotopic Evidence for the Diet of an Early Hominid, *Australopithecus africanus'*, *Science*, 283 (1999): 368.

Further reading

Jeremy Bernstein, *Plutonium* (New South, 2007)

Piers Bizony, *Atom* (Icon Books, 2008)

Brian Cathcart, *The Fly in the Cathedral* (Farrar, Straus and Giroux, 2004)

David M. Harland, *The Big Bang* (Springer, 2003)

Patrick W. Jackson, *The Chronologers' Quest* (Cambridge University Press, 2006)

Michael F. L'Annunziata, *Radioactivity* (Elsevier, 2007)

M. Levi, *On Nuclear Terrorism* (Harvard University Press, 2007)

Cherry Lewis, *The Dating Game* (Cambridge University Press, 2000)

Doug Macdougall, *Nature's Clocks* (University of California Press, 2008)

Harry Y. Mcsween, Jr, and Gary R. Huss, *Cosmochemistry* (Cambridge University Press, 2010)

S. F. Mason, *Chemical Evolution* (Clarendon Press, 1991)

Richard Muller, *Physics for Future Presidents* (Norton, 2008)

H. Reeves, Joel de Rosnay, Yves Coppens, and Dominique Simmonet, *Origins* (Arcade, 1996)

E. Segre, *From X-Rays to Quarks* (W. H. Freeman, 1980)

S. Singh, *Big Bang* (Fourth Estate, 2004)

C. Stringer, *The Origin of Our Species* (Allen Lane; Penguin Books, 2011)

C. Tuniz, R. Gillespie, and C. Jones, *The Bone Readers: Atoms, Genes and the Politics of Australia's Deep Past* (Sydney, Australia: Allen & Unwin; Left Coast Press, USA, 2009)

C. Tuniz, J. R. Bird, D. Fink, and G. Herzog, *Accelerator Mass Spectrometry* (CRC Press, LLC, USA, 1998)

P. Willmott, *An Introduction to Synchrotron Radiation* (Wiley, 2011)

M. Woolfson, *Time, Space, Stars and Man* (Imperial College Press, 2009)

T. Zoellner, *Uranium* (Penguin Books, 2009)

Radioactivity

Index

Available soon:

For more information visit our website
www.oup.com/vsi/

OXFORD
UNIVERSITY PRESS

Great Clarendon Street, Oxford, OX2 6DP,
United Kingdom

Oxford University Press is a department of the University of Oxford.
It furthers the University's objective of excellence in research, scholarship,
and education by publishing worldwide. Oxford is a registered trade mark of
Oxford University Press in the UK and in certain other countries

© Claudio Tuniz 2012

The moral rights of the author have been asserted

First Edition published in 2012

Impression: 1

British Library Cataloguing in Publication Data
Data available

Library of Congress Cataloging in Publication Data
Data available

ISBN 978-0-19-969242-2

Printed in Great Britain by
Ashford Colour Press Ltd, Gosport, Hampshire